Advanced Materials, Structures and Processing Technologies Based on Pulsed Laser

Advanced Materials, Structures and Processing Technologies Based on Pulsed Laser

Editors

Youmin Rong
Congyi Wu
Yu Huang

MDPI • Basel • Beijing • Wuhan • Barcelona • Belgrade • Manchester • Tokyo • Cluj • Tianjin

Editors
Youmin Rong
School of Mechanical Science and Engineering
Huazhong University of Science and Technology
Wuhan
China

Congyi Wu
School of Mechanical Science and Engineering
Huazhong University of Science and Technology
Wuhan
China

Yu Huang
School of Mechanical Science and Engineering
Huazhong University of Science and Technology
Wuhan
China

Editorial Office
MDPI
St. Alban-Anlage 66
4052 Basel, Switzerland

This is a reprint of articles from the Special Issue published online in the open access journal *Micromachines* (ISSN 2072-666X) (available at: www.mdpi.com/journal/micromachines/special_issues/advanced_pulsed_laser).

For citation purposes, cite each article independently as indicated on the article page online and as indicated below:

LastName, A.A.; LastName, B.B.; LastName, C.C. Article Title. *Journal Name* **Year**, *Volume Number*, Page Range.

ISBN 978-3-0365-2703-1 (Hbk)
ISBN 978-3-0365-2702-4 (PDF)

Contents

Editorial

Editorial for the Special Issue on Advanced Materials, Structures and Processing Technologies Based on Pulsed Laser

Youmin Rong [1,2,*], Congyi Wu [1,2] and Yu Huang [1,2,*]

1 State Key Lab of Digital Manufacturing Equipment and Technology, Huazhong University of Science and Technology, Wuhan 430074, China; cyw@hust.edu.cn
2 School of Mechanical Science and Engineering, Huazhong University of Science and Technology, Wuhan 430074, China
* Correspondence: rym@hust.edu.cn (Y.R.); yuhuang_hust@hust.edu.cn (Y.H.)

Citation: Rong, Y.; Wu, C.; Huang, Y. Editorial for the Special Issue on Advanced Materials, Structures and Processing Technologies Based on Pulsed Laser. *Micromachines* **2021**, *12*, 1507. https://doi.org/10.3390/mi12121507

Received: 26 November 2021
Accepted: 27 November 2021
Published: 1 December 2021

Pulsed lasers are lasers with a single laser pulse width of less than 0.25 s, operating only once in every certain time interval. Commonly used pulsed lasers are nanosecond, femtosecond, and picosecond lasers. A pulsed laser produces short pulses with a short interaction time with the material, which can largely avoid impact on the thermal movement of molecules and has a minimal thermal impact on the surrounding materials, thus, having significant advantages in precision microfabrication. It is now widely used in flexible electronics, chips, medicine, and other fields, such as photographic resin curing, microwelding, vision correction, heart stent manufacturing, etc. However, as an emerging processing technology, the application prospects of pulsed lasers have yet to be fully expanded, and there is still a need to continuously explore the mechanisms of interaction with materials, to manufacture advanced functional structures, and to develop advanced process technologies.

There are six papers published in this Special Issue focusing on pulsed laser machining. From metallic [1,2], inorganic [3] and polymeric [4,5] materials to biomass materials [6], from surface texture microstructures [1] and grain microstructures [2] to various types of functional contour structures [4,5], from the laser processing of metallic materials [1,2] and inorganic nonmetallic materials [3] to polymeric materials [4,5], from flexible electronics [4,5] and new energy batteries [1] to biomedicine [6], the articles in this Special Issue explore the specific applications of pulsed lasers in various fields.

In pulsed laser processing of metallic materials, Berhe et al. [1] studied the wettability behavior of structured and unstructured LiFePO4 electrodes. Firstly, the wettability morphology of the structured electrode was analyzed, and the electrode geometry as quantified in terms of ablation top and bottom width, ablation depth, and aspect ratio. From the result of the geometry analysis, the minimum measured values of aspect ratio and ablation depth were used as structured electrodes. Laser structuring with pitch distances of 112 μm, 224 μm, and 448 μm was applied. Secondly, the wettability of the electrodes was measured mainly by total wetting time and electrolyte spreading area. This study demonstrates that the laser-based structuring of the electrode increases the electrochemically active surface area of the electrode. The electrode structured with 112 μm pitch distance exhibited the fastest wetting at a time of 13.5 s. However, the unstructured electrode exhibited full wetting at a time of 84 s. Further, Fu et al. [2] studied the microstructure of a pulsed laser beam welded oxide dispersion-strengthened (ODS) eurofer steel. With a laser power of 2500 W and a duration of more than 3 ms, full penetration could be obtained in a 1 mm thick plate. Material loss was observed in the fusion zone due to metal vaporization, which can be fully compensated by the use of filler material. The solidified fusion zone consisted of an elongated dual phase microstructure with a bimodal grain size distribution. Nano-oxide particles were found to be dispersed in the steel. Electron backscattered diffraction (EBSD) analysis showed that the microstructure of the heat-treated joint was recovered with a substantially unaltered grain size and lower misorientations in different regions. The experimental results indicate that joints with fine grains and dispersed nano-oxide particles can be achieved via pulsed laser beam welding using a filler material and a postheat treatment.

In the pulsed laser processing of inorganic nonmetallic materials, Wang et al. [3] presented a facile laser-based surface texturing technique to modulate and control the surface functionalities (i.e., wettability and hardness) of Zr-based BMG. Laser surface texturing was first utilized to create periodic surface structures, and heat treatment was subsequently employed to control the surface chemistry. The experimental results indicated that the laser textured BMG surface became superhydrophilic immediately upon laser texturing, and turned superhydrophobic after heat treatment. Through surface morphology and chemistry analyses, it was confirmed that the wettability transition could be ascribed to the combined effects of the laser-induced periodic surface structure and controllable surface chemistry. In the meantime, the microhardness of the BMG surface had been remarkably increased as a result of laser surface texturing. The facile laser-based technique developed in this work has shown its effectiveness in modification and control of the surface functionalities for BMG, and it is expected to endow more useful applications.

In the pulsed laser processing of polymeric materials, Xu et al.'s [4] studies on the thermoplastic PET film and on temperature field-assisted ultraviolet nanosecond (UV–ns) pulse laser processing of polyethylene terephthalate (PET) film were performed to investigate the photothermal ablation mechanism and the effects of temperature on laser processing. The results showed that the UV–ns laser processing of PET film was dominated by the photothermal process in which PET polymer chains decomposed, melted, recomposed and reacted with the ambient gases. The ambient temperature changed the heat transfer and temperature distribution in the laser processing. Low ambient temperature reduced the thermal effect and an increase in ambient temperature improved its efficiency. Following on further, Wu et al. [5] prepared a contact spacer for a flexible tactile sensor using a thermoset polymer PI film UV–nanosecond laser cutting process. Taking a three factor, five level orthogonal experiment, the optimum laser cutting process was obtained (pulse repetition frequency 190 kHz, cutting speed 40 mm/s, and RNC 3). With the optimal process parameters, the minimum diameter was 24.3 ± 2.3 μm, and the minimum HAZ was 1.8 ± 1.1 μm. By analyzing the interaction process between nanosecond UV laser and PI film, the heating–carbonization mechanism was determined, and the influence of the process parameters on the quality of a micro-hole was discussed in detail in combination with this mechanism. It provides a new approach for the quantitative industrial fabrication of contact spacers in tactile sensors.

In the pulsed laser processing of biomass materials, Michalik et al. [6] deals with the medical application of diode-lasers. A short review of medical therapies is presented, considering the wavelength applied, continuous wave (cw) or pulsed regimes, and their therapeutic effects. Special attention was paid to the laryngological application of a pulsed diode laser with wavelength 810 nm, and the dermatologic applications of a 975 nm laser working at cw and pulsed mode. The efficacy of the laser procedures and a comparison of the pulsed and cw regimes is presented and discussed.

I would like to take this opportunity to thank all the authors for submitting their papers to this Special Issue. I would also like to thank all the reviewers for dedicating their time and helping to improve the quality of the submitted papers.

Funding: This work was supported by the National Natural Science Foundation of China (52005206, 51905191), the China Postdoctoral Science Foundation (2020TQ0110), Ministry of Industry and Information Technology's special project for high-quality development of manufacturing industry (TC200H02H), Guangdong HUST Industrial Technology Research Institute, Guangdong Provincial Key Laboratory of Manufacturing Equipment Digitization (2020B1212060014), Guangdong Basic and Applied Basic Research Foundation (2020A1515011393).

Conflicts of Interest: The author declares no conflict of interest.

References

1. Berhe, M.G.; Lee, D. A Comparative Study on the Wettability of Unstructured and Structured LiFePO4 with Nanosecond Pulsed Fiber Laser. *Micromachines* **2021**, *12*, 582. [CrossRef] [PubMed]
2. Fu, J.; Richardson, I.; Hermans, M. Microstructure Study of Pulsed Laser Beam Welded Oxide Dispersion-Strengthened (ODS) Eurofer Steel. *Micromachines* **2021**, *12*, 629. [CrossRef] [PubMed]
3. Wang, Q.; Cheng, Y.; Zhu, Z.; Xiang, N.; Wang, H. Modulation and Control of Wettability and Hardness of Zr-Based Metallic Glass via Facile Laser Surface Texturing. *Micromachines* **2021**, *12*, 1322. [CrossRef] [PubMed]
4. Xu, J.; Rong, Y.; Liu, W.; Zhang, T.; Xin, G.; Huang, Y.; Wu, C. Temperature Field-Assisted Ultraviolet Nanosecond Pulse Laser Processing of Polyethylene Terephthalate (PET) Film. *Micromachines* **2021**, *12*, 1356. [CrossRef] [PubMed]
5. Wu, C.; Zhang, T.; Huang, Y.; Rong, Y. PI Film Laser Micro-Cutting for Quantitative Manufacturing of Contact Spacer in Flexible Tactile Sensor. *Micromachines* **2021**, *12*, 908. [CrossRef] [PubMed]
6. Michalik, M.; Szymańczyk, J.; Stajnke, M.; Ochrymiuk, T.; Cenian, A. Medical Applications of Diode Lasers: Pulsed versus Continuous Wave (cw) Regime. *Micromachines* **2021**, *12*, 710. [CrossRef] [PubMed]

 *micromachines*

MDPI

Article

Temperature Field-Assisted Ultraviolet Nanosecond Pulse Laser Processing of Polyethylene Terephthalate (PET) Film

Jun Xu [1,2], Youmin Rong [1,2], Weinan Liu [1,2], Tian Zhang [1,2], Guoqiang Xin [1,2], Yu Huang [1,2] and Congyi Wu [1,2,*]

1 State Key Lab of Digital Manufacturing Equipment and Technology, Huazhong University of Science and Technology (HUST), Wuhan 430074, China; d202080271@hust.edu.cn (J.X.); rym@hust.edu.cn (Y.R.); wnliu@hust.edu.cn (W.L.); tianz@hust.edu.cn (T.Z.); gqxin@hust.edu.cn (G.X.); yuhuang_hust@hust.edu.cn (Y.H.)

2 School of Mechanical Science and Engineering, Huazhong University of Science and Technology (HUST), Wuhan 430074, China

* Correspondence: cyw@hust.edu.cn

Abstract: Understanding the mechanism of and how to improve the laser processing of polymer films have been important issues since the advent of the procedure. Due to the important role of a photothermal mechanism in the laser ablation of polymer films, especially in transparent polymer films, it is both important and effective to adjust the evolution of heat and temperature in time and space during laser processing by simply adjusting the ambient environment so as to improve and understand the mechanism of this procedure. In this work, studies on the pyrolysis of PET film and on temperature field-assisted ultraviolet nanosecond (UV-ns) pulse laser processing of polyethylene terephthalate (PET) film were performed to investigate the photothermal ablation mechanism and the effects of temperature on laser processing. The results showed that the UV-ns laser processing of PET film was dominated by the photothermal process, in which PET polymer chains decomposed, melted, recomposed and reacted with the ambient gases. The ambient temperature changed the heat transfer and temperature distribution in the laser processing. Low ambient temperature reduced the thermal effect and an increase in ambient temperature improved its efficiency (kerf width: 39.63 μm at −25 °C; 48.30 μm at 0 °C; 45.81 μm at 25 °C; 100.70 μm at 100 °C) but exacerbated the thermal effect.

Keywords: laser processing; PET film; transparent polymer; temperature field; ultraviolet nanosecond pulse laser; laser photothermal ablation

Citation: Xu, J.; Rong, Y.; Liu, W.; Zhang, T.; Xin, G.; Huang, Y.; Wu, C. Temperature Field-Assisted Ultraviolet Nanosecond Pulse Laser Processing of Polyethylene Terephthalate (PET) Film. *Micromachines* **2021**, *12*, 1356. https://doi.org/10.3390/mi12111356

Academic Editor: Ioanna Zergioti

Received: 30 September 2021
Accepted: 30 October 2021
Published: 2 November 2021

1. Introduction

Polymer films are widely applied in electronic devices [1], flexible devices [2], energy devices [3], etc. [4,5], due to their unique physical and chemical properties. Thus, high-precision processing methods for polymer films have received increasing attention. In particular, the rapid development of large-scale integrated circuits and various functional elements has created higher requirements for the processing precision, quality and efficiency of polymer films. Laser processing, as a non-contact processing method, has received growing attention due to its adaptability, low divergence and ability to render the selective removal of polymers with high spatial resolution [6,7], and it is widely applied in the processing of a variety of materials such as metals, glasses, polymers, etc. [8–12] Therefore, the high-precision laser processing of polymer films has attracted much attention due to its unique advantages [12]. In past decades, many polymers have been studied for laser micromachining such as polyethylene terephthalate (PET) [13], polyimide (PI) [14], polydimethylsiloxane (PDMS) [15], polymethyl methacrylate (PMMA) [16], polytetrafluoroethylene (PTFE) [17], etc. [4,18] However, questions regarding the improvement of the precision, quality and efficiency of laser processing, as well as the laser ablation process and mechanism, still prevail.

According to the principles of laser processing and previous research,, there are three ways to improve the procedure by changing the interactions between lasers and materials: (1) adjusting the parameters of lasers, such as power, wavelength, pulse width, etc. [19,20] (2) adjusting the physical or chemical properties of materials, such as absorption coefficient, reflectivity, thermal conductivity, etc. [21–23] (3) adding physical fields or changing the external chemical environment such as temperature field, electromagnetic field, supersonic wave or specific chemical solutions [24,25]. Adjusting the laser parameters is an effective method to improve processing quality, but its effectiveness is limited in certain materials. Furthermore, although changing the material properties can bring some improvement, usually this change is irreversible and difficult for the formed polymer films. Compared with the above two methods, adding physical fields or changing the external chemical environment is more simple, effective and widely applicable.

The laser ablation process of polymer films is a combination of photothermal and photochemical processes [26]. Between these two mechanisms, the photothermal mechanism plays an important role in laser processing, especially for transparent polymer films with poor absorption. Therefore, the absorption, conversion and transfer of heat, as well as the change and distribution of temperature, affect the ablation and decomposition of polymer films during laser processing in important ways. This is key to correctly managing the evolution of heat and temperature in time and space during the laser processing of polymer films via simple adjustments to the ambient environment, so as to improve the procedure. It also helps to understand the process and mechanism of the laser processing of polymer films.

In this study, an external temperature field was applied to assist the laser processing of polymer films to investigate the effects of ambient temperature on the procedure, which made the laser ablation mechanism clearer and more understandable. The selected experimental material was PET film, which is widely used in both the electronics industry and daily life [27,28]. The applied laser was an ultraviolet (355 nm) nanosecond (UV-ns) pulse laser due to it being less expensive and possessing an appropriate wavelength for polymer. Firstly, thermogravimetric analysis (TGA), thermogravimetric analysis, Fourier transform infrared spectroscopy (TGA-FTIR), pyrolysis–gas chromatography-mass spectrometry (GC-MS) and Raman spectrometry were performed to study the thermal decomposition process and laser photothermal ablation mechanism of PET films. Secondly, laser processing experiments (single-spot ablation and line ablation) of PET films at various ambient temperatures (−25 °C, 0 °C, 50 °C, 75 °C, 100 °C) were performed to investigate the effects of ambient temperature on the morphology and geometric characteristics (size, processing area, heat affected zone (HAZ)) of features (hole and kerf) so as to study the influencing mechanism of ambient temperature on the laser processing of PET film. The results showed that the laser processing of PET film was dominated by the photothermal decomposition process, in which PET polymer chains decomposed, melted, recomposed and reacted with the ambient gases. Changes in ambient temperature affected the laser processing of PET film: an increase in ambient temperature changed the heat transfer and temperature distribution during the laser processing, while a lower ambient temperature reduced the thermal effect. Increasing ambient temperature also improved the efficiency (kerf width: 39.63 μm at −25 °C; 45.81 μm at 25 °C; 100.70 μm at 100 °C) but exacerbated the thermal effect. This work provides effective methods to study the laser processing mechanism of polymer films, as well as an approach to improve this procedure.

2. Materials and Methods

2.1. Materials

The raw material that was processed was commercial transparent PET film (Thickness: 0.2 μm), supplied by Xilu Photoelectricity Technic CO., LTD (Shenzhen, China).

2.2. Laser Processing System

The laser processing system used was an experimental platform built by the team. As shown in Figure 1a, the laser processing system consisted of an ultraviolet nanosecond (UV-ns) laser (Poplar-355-15A5, Wuhan Huaray Precision Laser Co. Ltd., Wuhan, China), two reflectors, a three-dimensional scanning galvanometer (intelliSCAN 14, Scanlab, Puchheim, Germany), a beam expander, a proportion–integral–differential (PID) temperature control platform and a computer. The output power of the laser was controlled by adjusting laser repetition frequency and measured with a power meter. The specific parameters of the laser processing system are listed in Table 1. PET film samples were placed on the PID temperature control platform, the surface temperature of which was regulated by either ceramic heater heating or liquid nitrogen cooling. UV-ns laser, scanning galvanometer and the PID temperature control platform were controlled by computer so that laser processing parameters (laser power, scanning speed and ambient temperature) could be adjusted.

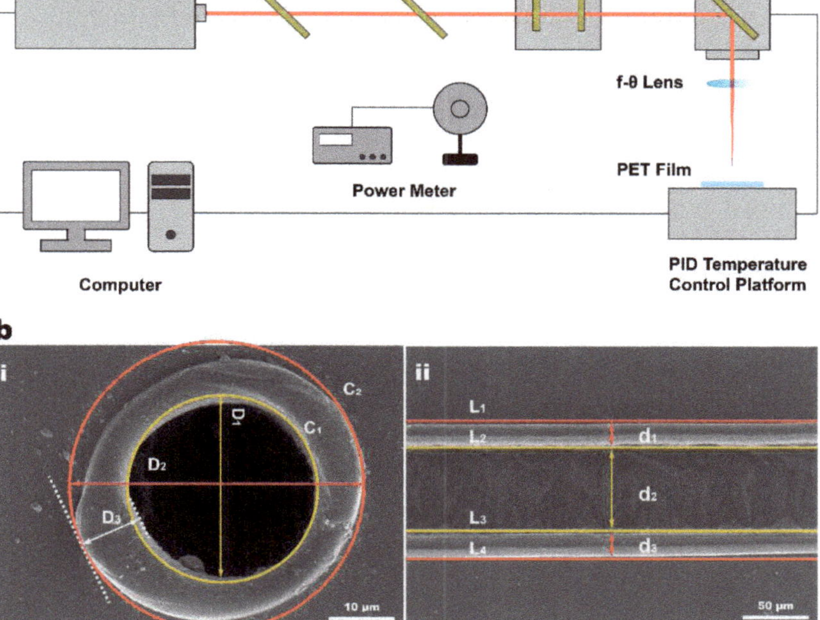

Figure 1. The Schematic diagram of the laser processing of PET film (**a**), and the processing quality evaluation (**b**) of different features (hole (i) and kerf(ii)).

Table 1. The specific parameters of laser processing system.

Parameters	Values
Wavelength	355 nm
Pulse width	16 ± 2 ns@50 kHz
Beam diameter	>11.8 µm
Focal length	167 mm

2.3. Experimental Design

Photothermal ablation and thermal effects play important roles in laser processing. Heat generation and transfer can be regulated by adjusting processing parameters (laser power, scanning speed, scanning times, etc.), material properties or environmental conditions. In our experimental design, in order to investigate the influences of ambient temperature on heat generation and transfer during laser processing, single-point and line laser processing were performed. In these experiments, processing parameters (laser power, repetition frequency, duration time, scanning times, etc.) were constant and the effects of ambient temperature were investigated by temperature gradient experiments. Table 2 shows the levels of ambient temperature and processing parameters in the single-point and line laser processing experiments.

Table 2. The experimental design.

Experiment	No.	Temperature (°C)	Parameters
Single-Point Laser Processing	1	−25	Laser Power = 0.9 W Repetition Frequency = 200 kHz Duration Time = 0.05s
	2	0	
	3	25	
	4	50	
	5	75	
	6	100	
Line Laser Processing	1	−25	Laser Power = 0.9 W Repetition Frequency = 200 kHz Scanning Speed = 50 mm/s Scanning Times = 30
	2	0	
	3	25	
	4	50	
	5	75	
	6	100	

Based on the typical features (hole and kerf) of PET film samples after laser processing, processing quality evaluation methods were established. As shown in Figure 1b, in the single-point laser processing experiment, edge profiles of ablation hole circles C_1 with diameter D_1 and C_2 with diameter D_2 were clearly observed. HAZ was located between C_1 and C_2 with spacing of D_3, the bump formed after PET melting and solidification. In the line laser processing experiment, edge profiles of ablation kerf lines L_1, L_2, L_3 and L_4 could be clearly observed. Among them, HAZ above kerf was located between L_1 and L_2 with spacing of d_1. Kerf was located between L_2 and L_3 with spacing of d_2. HAZ below kerf was located between L_3 and L_4 with spacing of d_3. The values of hole diameter (D_H), processing area diameter of hole (D_{PH}), HAZ width of hole or bump (D_{HH}, maximum) and kerf width (W_K) were initialized as the values of D_1, D_2, D_3 and d_2, respectively. HAZ width of kerf (W_{HK}) and processing area diameter of kerf (W_{PK}) were calculated using the following equations:

$$W_{HK} = MAX:(d_1, d_3) \tag{1}$$

$$W_{PK} = d_1 + d_2 + d_3 \tag{2}$$

2.4. Characterizations

A scanning electron microscope (SEM, HITACHI SU3900, Tokyo, Japan) was applied to observe surface morphologies of the kerf and hole of PET film samples. A thermogravimetry-infrared association meter (TGA-FTIR, PerkinElmer, Waltham, MA, USA) was used for the thermogravimetric analysis (TGA, 25–1000 °C) of PET film, and IR spectra (500–4000 cm^{-1}) of gaseous decomposition products of PET film at different temperatures were collected during thermal decomposition process. A pyrolysis–gas chromatography-mass spectrometer (GC-MS, Agilent 7890A/5975C, Santa Clara, CA, USA) was used to analyze gaseous decomposition products of PET film during rapid pyrolysis (25–1000 °C, 50 °C/ms, N_2). A laser confocal Raman spectrometer (532 nm, 500–3000 cm^{-1},

LabRAM HR800, Horiba JobinYvon, Palaiseau, France) was used for the Raman spectra of PET film and its solid decomposition products.

3. Results and Discussion

3.1. Laser Ablation Mechanism of PET Film

During the laser processing of polymers, laser ablation is a combination of photothermal and photochemical processes [26,29]. Between these two mechanisms, the photothermal mechanism plays an important role in laser processing, especially for transparent polymers with poor absorption. It is important and effective to analyze the laser ablation products during photothermal ablation in order to understand how the laser ablation of polymers works, as well as improving its quality. Therefore, thermogravimetric analysis was performed to analyze the thermal decomposition process of the PET film. Figure 2 shows the TGA-DTG curves of PET film. The thermal decomposition process of the PET film consisted of two steps. First, when the temperature reached about 310.5 °C, the PET film started to decompose and its weight dropped to about 40.7%. Second, when the temperature reached about 475.0 °C, the PET film further decomposed and its residue weight was about 10.6%.

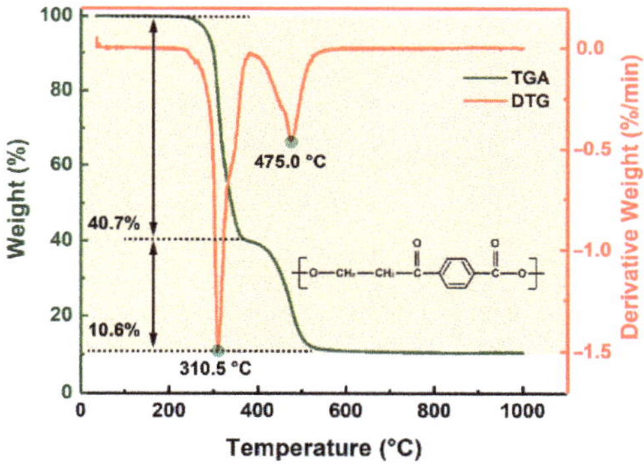

Figure 2. TGA-DTG curves of the PET film.

After this, pyrolysis–gas chromatography-mass spectrometry was performed to analyze the gaseous decomposition products of the PET film during pyrolysis. Figure 3 shows the ion current of evolved gases during its pyrolysis. The results showed that the gaseous decomposition products produced during the pyrolysis of the PET film were mainly composed of: carbon dioxide (3.90%); 1-phenyl-1,2-propanedione (4.65%); benzoic acid (12.56%); o-diacetylbenzene (3.90%); benzimidazole-2-carboxaldehyde (11.56%); 1-methyl-, oxime 4-acetylbenzoic acid (26.17%); 2,4,5-triphenyl-1,3-oxazole (11.67%); (4-dimethylamino-phenyl)-(4-nonyloxy-phenyl)-methanone (4.28%); etc. Table 3 lists the information about the gaseous decomposition products of the PET film in detail. These results showed that during the pyrolysis process, the PET film either decomposed into small, molecular-like carbon dioxide, or recomposed to form new substances such as 2,4,5-triphenyl-1,3-oxazole. Ambient gases also participated in the pyrolysis process of the PET film. During the laser processing of the PET film, the decomposition process and products of the PET film were more complicated, clearly due to the more complex environment and the more complex interactions between laser, heat and matter.

Table 3. Composition table of the evolved gaseous products during pyrolysis of PET film.

Peak	RT (min)	Area (%)	ID	CAS#	2D	3D
1#	1.249	3.90	Carbon dioxide	000124-38-9	O=C=O	
2#	8.757	4.65	1-Phenyl-1,2-propanedione	000579-07-7		
3#	9.778	12.56	Benzoic acid	000065-85-0		
4#	13.943	3.90	o-Diacetylbenzene	000704-00-7		
5#	14.602	11.56	Benzimidazole-2-carboxaldehyde, 1-methyl-, oxime	003013-07-8		
6#	15.369	26.17	4-Acetylbenzoic acid	000586-89-0		
7#	24.439	11.67	2,4,5-triphenyl-1,3-oxazole	000573-34-2		
8#	27.605	4.28	(4-Dimethylamino-phenyl)-(4-nonyloxy-phenyl)-methanone	300382-52-9		

Furthermore, TGA-FTIR was used to analyze gaseous decomposition products formed during the pyrolysis of the PET film. FTIR spectra of gaseous products at different temperatures from 50 to 1000 °C are presented in Figures 4 and 5. There was no obvious absorption peak below about 300 °C, which meant no decomposition. Above 300 °C, obvious absorption peaks gradually appeared and the intensity reached its maximum at

325 °C and 489 °C (Figure 5a), respectively, which indicated that the PET film began to decompose gradually and was consistent with the thermal analysis results. As shown in Figure 5b, the characteristic peaks of carbon dioxide (674 cm^{-1} and 2342 cm^{-1}), phenyl (1192 cm^{-1} for C-H, 1486 cm^{-1} for C=C) and carbonyl group (1730 cm^{-1} for C=O) were observed in the gaseous products at 325 °C, which might be due to the formation of small molecules such as carbon dioxide and benzoic acid. The peaks at 2822 cm^{-1} and 2954 cm^{-1} were assigned to CH_x stretching vibrations [30,31]. At 489 °C, a stronger C-H characteristic peak was observed at 2936 cm^{-1}. When the temperature rose above 996 °C, the PET film further decomposed and recomposed into small molecules. Here, the characteristic peaks of carbon nitrogen compounds (1132 cm^{-1} for C-N, 2108 cm^{-1}, 2182 cm^{-1} and 2360 cm^{-1} for C≡N) and carbon oxides (1342 cm^{-1} for C-O) appeared, but the characteristic peaks of phenyl and carboxyl groups did not, which indicated the PET film not only decomposed, but also recomposed and even reacted with ambient gases under high temperature. These FTIR results at different temperatures were consistent with the pyrolysis results.

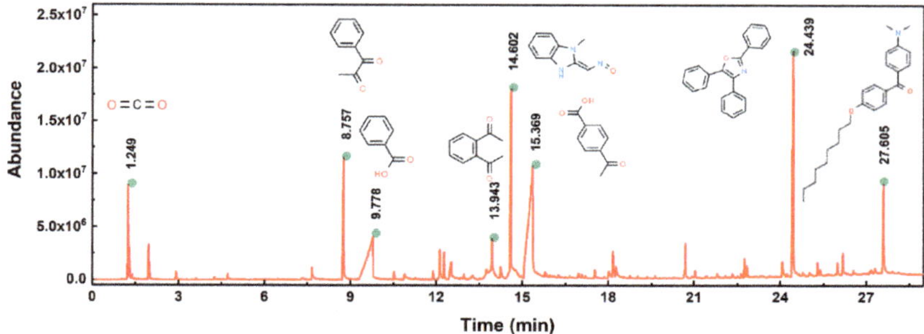

Figure 3. Ion current of evolved gaseous products during pyrolysis of PET film.

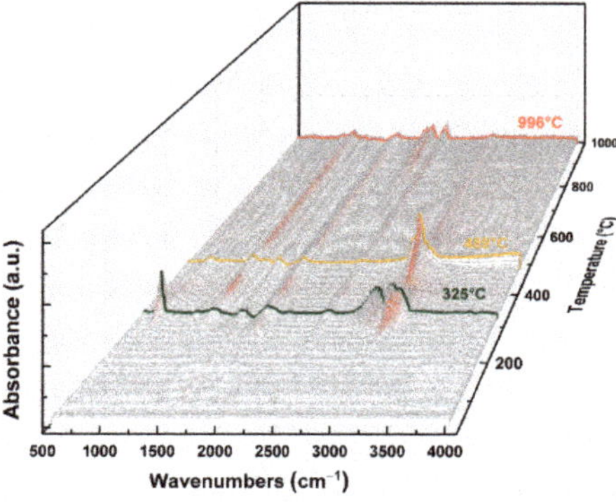

Figure 4. Infrared spectrogram of evolved gaseous products during pyrolysis of PET film.

In addition to the gaseous decomposition products, the solid residues after laser ablation were analyzed by a Raman spectrometer. Figure 6 shows the Raman spectra of different areas of the PET film after laser processing: the unprocessed area of the native PET film (Point 1); the outer edge of the bump (Point 2); the surface of the bump (Point 3); and the inner surface of the kerf, or inner edge of the bump (Point 4). The Raman spectrum of the PET film was consistent with previous reports [32], in which characteristic peaks of terephthalate and of ethylene glycol (CH_2-CH_2-O) functional groups were found. In detail, the characteristic peaks at 632 cm^{-1} (CCC in plane bending (phenyl)), 796 cm^{-1} (C-H out of plane bending (phenyl)), 860 cm^{-1} (C-C stretching (phenyl breathing), C-O stretching), 1095 cm^{-1} (C-C stretching (glycol)), 1118 cm^{-1} (C-H in plane bending (phenyl), C-O stretching), 1187 cm^{-1} (C-H in plane bending (phenyl)), 1292 cm^{-1} (C-C stretching (phenyl), C-O stretching), 1416 cm^{-1} (C-C stretching (phenyl)), 1615 cm^{-1} (C=C stretching (phenyl)) and 1726 cm^{-1} (C=O stretching) were observed in the Raman spectrum of the PET film (Point 1) [32]. After laser ablation, a kerf, a bump and splashes were observed in the laser processing area due to the melting of the PET film caused by heat. At Points 2, 3 and 4, the characteristic peak intensity of the functional groups in the PET film decreased to varying degrees, and the closer these were to the central laser processing area, the lower their intensity, and in some cases they even vanished. This indicated that during the laser processing of PET film, the polymer chains absorbed enough heat to cause their bonds to break and decompose into small molecular fragments, due to the photothermal mechanism. The closer to the processing center, the higher the temperature and the more complete the decomposition of the polymer chains was.

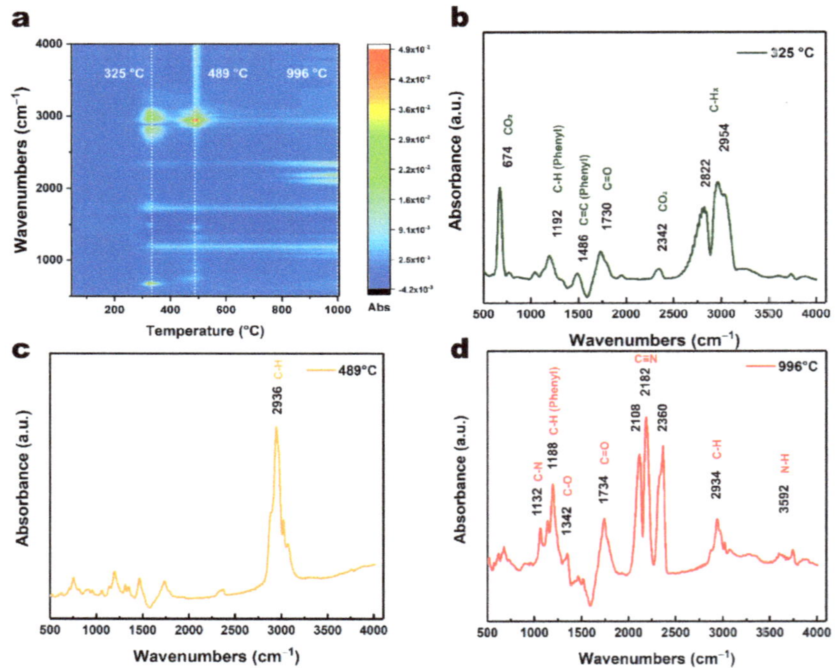

Figure 5. Infrared spectral intensity maps of evolved gaseous products during pyrolysis of PET film (**a**), and infrared spectrogram at typical temperatures: (**b**) 335 °C, (**c**) 489 °C and (**d**) 996 °C.

According to the above results, showing gaseous products caused by thermal decomposition and solid products caused by laser ablation of the PET film, the laser processing process and mechanism of the PET film were clear. Figure 7 illustrates the schematic diagram of the interaction mechanism between nanosecond UV laser and the PET film during laser processing. Firstly, the PET film was almost transparent at the laser excitation wavelength (355 nm), and the absorbed single photon energy (3.49 eV at 355 nm) was insufficient to break the polymer backbone bonds directly (3.69 eV for C-C). The photothermal mechanism dominated the laser processing or ablation process of the PET film, and the decomposition of the PET polymer chains was mainly pyrolysis [13,26,33–35]. Secondly, regarding the laser irradiation and photothermal conversion, the PET polymer chains either decomposed into small molecules or short-chain polymers, or recomposed [13,26,36]. At the same time, ambient gases such as oxygen and nitrogen also participated in this process. Thirdly, as the temperature changed, thermoplastic PET film melted and solidified, resulting in the formation of bumps around the kerf [13], around which the splatter deposited.

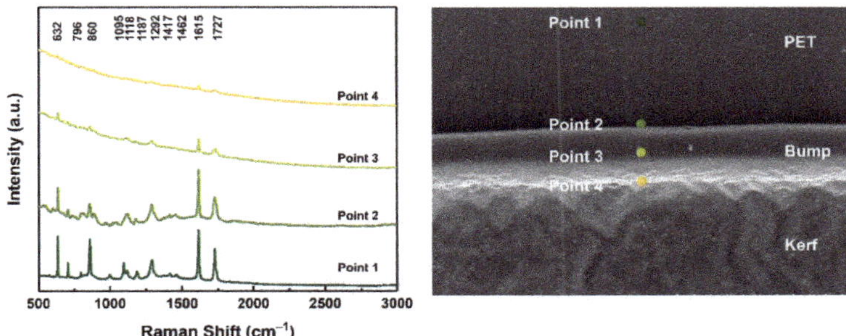

Figure 6. The Raman spectra of PET film after laser processing.

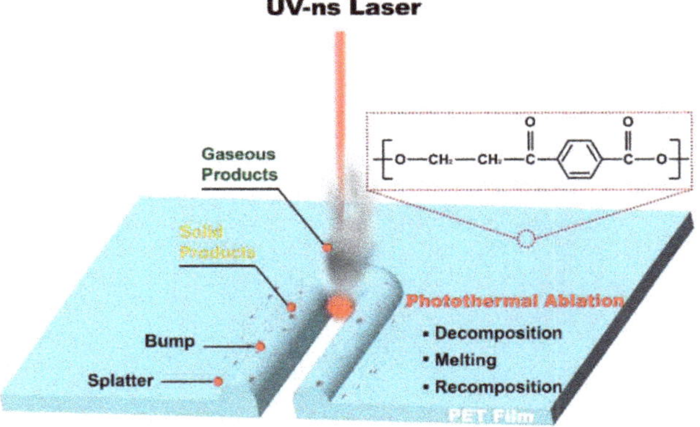

Figure 7. Schematic diagram of the interaction mechanism between the UV-ns laser and the PET film.

3.2. The Morphology of Features at Different Temperatures

According to the laser processing mechanism of PET film, it can be seen that the conversion, absorption and transfer of heat as well as the change in temperature had vital influences on this process. Therefore, the external temperature field-assisted laser processing of PET film experiment was designed and performed to investigate the laser processing of PET film at different ambient temperatures and the effects of temperature on the laser processing of PET film, in which single-point laser ablation and line laser ablation were carried out at different temperatures (−25 °C, 0 °C, 25 °C, 50 °C, 75 °C, 100 °C).

Figures 8 and 9 show the morphologies and sizes of ablation holes in the single-point laser ablation of PET films at different temperatures. As the ambient temperature increased, the so too did the diameter of the ablation hole and processing area. At room temperature (25 °C), the diameter of the ablation hole and processing area was 29.36 µm and 45.4 µm, respectively. When ambient temperature decreased to 0 °C and −25 °C, the diameter of the ablation hole was 21.67 µm and 20.92 µm, respectively, and the diameter of processing area was 42.85 µm and 36.30 µm, respectively. When the ambient temperature was above room temperature, the diameter of the ablation hole was 30.42 µm at 50 °C, 30.26 µm at 75 °C and 24.84 µm at 100 °C, respectively, and the diameter of processing area was 47.60 µm at 50 °C, 49.57 µm at 75 °C and 47.88 µm at 100 °C, respectively. As for the HAZ or the bump, the size was 10.12 µm at −25 °C, 9.77 µm at 0 °C, 12.37 µm at 25 °C, 10.29 µm at 50 °C, 10.11 µm at 75 °C and 12.37 µm at 100 °C, respectively. The effect of ambient temperature on the size of the HAZ was not obvious. In addition to the change in feature sizes, the morphology of the holes also changed. When the temperature was relatively low (−25 °C, 0 °C, 25 °C), the holes were smooth and clean. However, when the temperature increased (50 °C, 75 °C, 100 °C), wrinkles and debris appeared.

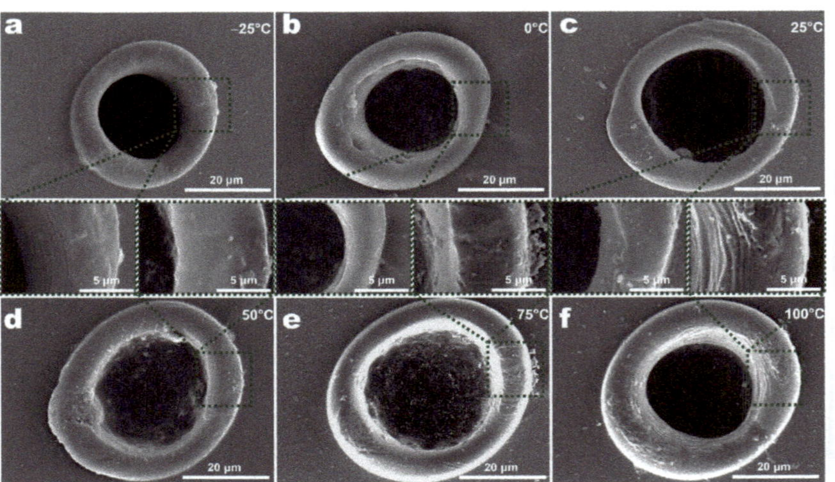

Figure 8. SEM images of holes of PET films at different ambient temperatures in the single-spot laser processing experiment: (**a**) −25 °C, (**b**) 0 °C, (**c**) 25 °C, (**d**) 50 °C, (**e**) 75 °C, (**f**) 100 °C.

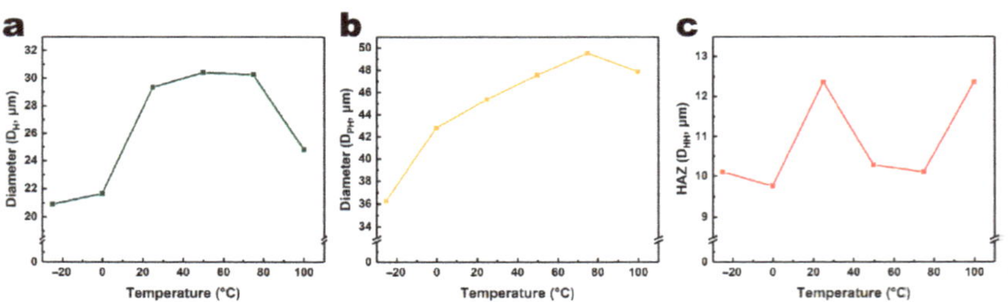

Figure 9. Hole diameter (**a**), processing area diameter (**b**) and HAZ (**c**) of PET films at different ambient temperature in the single spot laser processing experiment.

Figures 10 and 11 show the morphologies and sizes of ablation kerfs in the line laser ablation of PET films at different ambient temperatures. Similar to the single-spot ablation experiment, as the ambient temperature increased, the width of the ablation kerf and processing area also increased. The kerf width was 39.63 µm at −25 °C, 48.30 µm at 0 °C, 45.81 µm at 25 °C, 69.66 µm at 50 °C, 77.36 µm at 75 °C and 100.70 µm at 100 °C. The processing area width was 71.96 µm at −25 °C, 79.84 µm at 0 °C, 77.43 µm at 25 °C, 104.52 µm at 50 °C, 115.30 µm at 75 °C and 134.73 µm at 100 °C. The HAZ width also showed an approximate slow increase trend with increased ambient temperature. The team systematically studied the influences of laser processing parameters (repetition rate, cutting speed and cutting times) on laser processing of PET film [13]. Although optimizing laser processing parameters could improve laser processing to a certain extent, its effect was very limited. Compared to this, adjusting ambient temperature in the current study improved the laser processing process of the PET film more effectively and significantly, which means that this method possesses greater application potential and scope in actual industrial production.

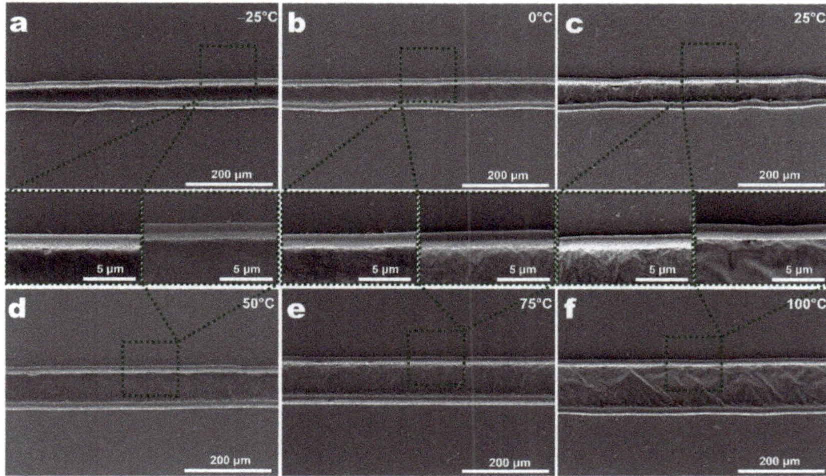

Figure 10. SEM images of kerfs of PET films at different ambient temperatures in line laser processing experiment: (**a**) −25 °C, (**b**) 0 °C, (**c**) 25 °C, (**d**) 50 °C, (**e**) 75 °C, (**f**) 100 °C.

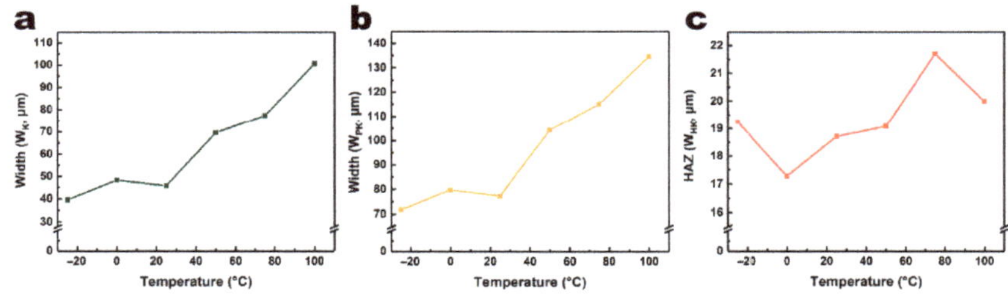

Figure 11. Kerf width (**a**), processing area width (**b**) and HAZ (**c**) of PET films at different ambient temperatures in line laser processing experiment.

3.3. The Effects of Temperature on Laser Processing of PET Film

Through the above analysis of the morphologies and sizes of the features (hole and kerf), the effects of ambient temperature on the laser processing of the PET film were revealed. Increasing the ambient temperature promoted the pyrolysis of the PET film and improved laser ablation efficiency, so the sizes of the hole, kerf and processing area increased with the increase in ambient temperature. However, when the ambient temperature was too high, this promotion and improvement might have caused the non-uniform and incomplete decomposition of the PET polymer chain, resulting in debris deposited in the processing area. Indeed, low ambient temperature reduced the HAZ and the thermal effect, while the increase in ambient temperature resulted in an increase in the HAZ and thermal stress, which caused the increase of size of the HAZ or bump and the appearance of wrinkles. The root cause of the above effects on the laser processing of the PET film was that the increase in ambient temperature promoted the heat transfer in the laser processing. The above analysis showed that a proper ambient temperature could not only improve the laser processing efficiency of the PET film, but also ensure good quality [36].

In order to verify and study the effects of ambient temperature on the laser processing of PET film further, the Raman spectra (at HAZ or Bump) of PET films processed at different temperatures were acquired, as shown in Figure 12. As the ambient temperature increased, the characteristic peaks of the functional groups and chemical bonds in the PET film gradually decreased or even vanished. When the ambient temperature was low (−25 °C, 0 °C, 25 °C), the peaks at 632 cm^{-1}, 860 cm^{-1}, 1187 cm^{-1}, 1292 cm^{-1}, etc., were recognizable. However, when the ambient temperature was above 50 °C, these peaks gradually decreased or even disappeared. The difference in Raman spectra of PET films processed at different ambient temperatures proved that the increase in ambient temperature promoted heat transfer during the laser processing, which promoted the pyrolysis of the PET film and made the HAZ increase.

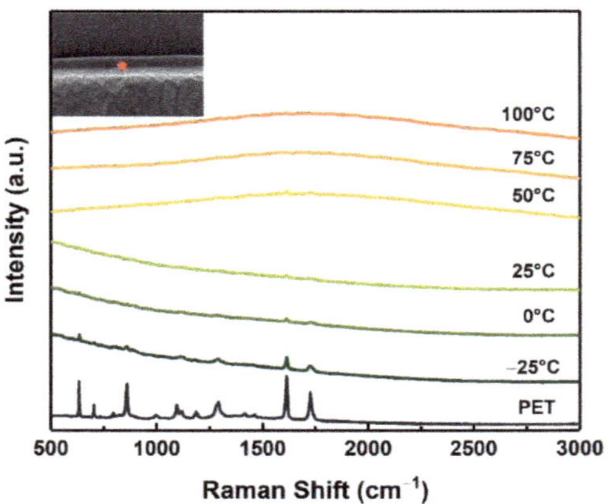

Figure 12. The Raman spectra (at HAZ or bump) of PET films at different temperatures.

4. Conclusions

The photothermal ablation mechanism and the effects of ambient temperature on the laser processing of PET film were investigated. According to our studies on the pyrolysis of PET film and the temperature field-assisted UV-ns pulse laser processing, the main conclusions are as follows:

1. When the PET film is almost transparent at the laser excitation wavelength (355 nm), and the single photon energy at 355 nm is insufficient to break the polymer backbone bonds directly, the laser processing of PET film is dominated by the photothermal decomposition process (pyrolysis).

2. During the laser processing of PET film, PET polymer chains decompose into small fragments, which recompose, and ambient gases also participate in this process. As the ambient temperature changes, thermoplastic PET film melts, resulting in the formation of a bump. Splatter is deposited in the processing area.

3. An adjustment in ambient temperature affects the laser processing of PET film. An increase in ambient temperature changes the heat transfer and temperature distribution in the laser processing. A low ambient temperature reduces the thermal effect, and an increase in ambient temperature improves the efficiency (kerf width: 39.63 µm at −25 °C, 45.81 µm at 25 °C, 100.70 µm at 100 °C) but exacerbates the thermal effect.

This work provides effective methods to study the laser processing mechanism of polymer films as well as an approach to improve the laser processing of polymer films.

Author Contributions: Conceptualization and methodology, C.W.; validation, T.Z.; formal analysis, G.X.; investigation, W.L.; data curation, writing—original draft preparation and writing—review and editing, J.X.; supervision and project administration, Y.R.; resources and funding acquisition, Y.H. All authors have read and agreed to the published version of the manuscript.

Funding: This work was supported by the National Natural Science Foundation of China (52005206, 51905191), the China Postdoctoral Science Foundation (2020TQ0110), Ministry of Industry and Information Technology's special project for high-quality development of manufacturing industry (TC200H02H), National Key R&D Program of China (Grant No. 2020YFB2007600), Scientific Research Plan Guiding Project of Hubei Province Education Department (B2019246), Guangdong HUST Industrial Technology Research Institute, Guangdong Provincial Key Laboratory of Digital Manufacturing Equipment (2020B1212060014), Guangdong Basic and Applied Basic Research Foundation (2020A1515011393).

Institutional Review Board Statement: Not applicable.

Informed Consent Statement: Not applicable.

Data Availability Statement: The data presented in this study are available on request from the corresponding author. The data are not publicly available because the data are also part of an ongoing study.

Acknowledgments: The general characterization facilities are provided by the Flexible Electronics Manufacturing Laboratory in Experiment Center for Advanced Manufacturing and Technology in School of Mechanical Science & Engineering of HUST.

Conflicts of Interest: The authors declare no conflict of interest.

References

1. Zang, Y.; Shen, H.; Huang, D.; Di, C.A.; Zhu, D. A dual-organic-transistor-based tactile-perception system with signal-processing functionality. *Adv. Mater.* **2017**, *29*, 1606088. [CrossRef] [PubMed]
2. Cheng, T.; Zhang, Y.; Lai, W.Y.; Huang, W. Stretchable thin-film electrodes for flexible electronics with high deformability and stretchability. *Adv. Mater.* **2015**, *27*, 3349–3376. [CrossRef]
3. Kim, M.P.; Ahn, C.W.; Lee, Y.; Kim, K.; Park, J.; Ko, H. Interfacial polarization-induced high-k polymer dielectric film for high-performance triboelectric devices. *Nano Energy* **2021**, *82*, 105697. [CrossRef]
4. Singha, S.S.; Khareb, A.; Joshi, S.N. Fabrication of microchannel on polycarbonate below the laser ablation threshold by repeated scan via the second harmonic of Q-switched Nd:YAG laser. *J. Mater. Process. Technol.* **2020**, *55*, 359–372. [CrossRef]
5. Dahotre, N.B.; Harimkar, S. *Laser Fabrication and Machining of Material*; Springer: Boston, MA, USA, 2001. [CrossRef]
6. Chryssolouris, G. *Laser Machining—Theory and Practice, Mechanical Engineering Series*; Springer: New York, NY, USA, 1991. [CrossRef]
7. Kamlage, G.; Bauer, T.; Ostendorf, A.; Chichkov, B.N. Deep drilling of metals by femtosecond laser pulses. *Appl. Phys. A.* **2003**, *77*, 307–310. [CrossRef]
8. Jia, W.; Yu, J.; Chai, L.; Wang, C.Y. Micromachining soda-lime glass by femtosecond laser pulses. *Front. Phys.* **2015**, *10*, 1–4. [CrossRef]
9. Frank, P.; Shaw-Stewart, J.; Lippert, T.; Boneberg, J.; Leiderer, P. Laser-induced ablation dynamics and flight of thin polymer films. *Appl. Phys. A* **2011**, *104*, 579–582. [CrossRef]
10. Jiang, H.; Ma, C.; Li, M.; Cao, Z. Femtosecond laser drilling of cylindrical holes for carbon fiber-reinforced polymer (CFRP) composites. *Molecules* **2021**, *26*, 2953. [CrossRef] [PubMed]
11. Livingstone, S.A.J.; Chua, K.L.; Black, I. Experimental development of a machining database for the CO_2 laser cutting of ceramic tile. *J. Laser Appl.* **1997**, *9*, 233–241. [CrossRef]
12. Wu, C.; Li, M.; Huang, Y.; Rong, Y. Cutting of polyethylene terephthalate (PET) film by 355 nm nanosecond laser. *Opt. Laser Technol.* **2021**, *133*, 106565. [CrossRef]
13. Ravi-Kumar, S.; Lies, B.; Lyu, H.; Qin, H. Laser ablation of polymers: A review. *Procedia Manuf.* **2019**, *34*, 316–327. [CrossRef]
14. Rong, Y.; Huang, Y.; Lin, C.; Liu, Y.; Shi, S.; Zhang, G.; Wu, C. Stretchability improvement of flexible electronics by laser micro-drilling array holes in PDMS film. *Opt. Lasers Eng.* **2020**, *134*, 106307. [CrossRef]
15. Beinhorn, F.; Ihlemann, J.; Luther, K.; Troe, J. Micro-lens arrays generated by UV laser irradiation of doped PMMA. *Appl. Phys. A* **1999**, *68*, 709–713. [CrossRef]
16. Yong, J.; Fang, Y.; Chen, F.; Huo, J.; Yang, Q.; Bian, H.; Du, G.; Hou, X. Femtosecond laser ablated durable superhydrophobic PTFE films with micro-through-holes for oil/water separation: Separating oil from water and corrosive solutions. *Appl. Surf. Sci.* **2016**, *389*, 1148–1155. [CrossRef]
17. Caiazzo, F.; Curcio, F.; Daurelio, G.; Minutolo, F.M.C. Laser cutting of different polymeric plastics (PE, PP and PC) by a CO_2 laser beam. *J. Mater. Process. Technol.* **2005**, *159*, 279–285. [CrossRef]
18. Schmidt, H.; Ihlemann, J.; Wolff-Rottke, B.; Luther, K.; Troe, J. Ultraviolet laser ablation of polymers: Spot size, pulse duration, and plume attenuation effects explained. *J. Appl. Phys.* **1998**, *83*, 5458–5468. [CrossRef]

19. Stankova, N.E.; Atanasov, P.A.; Nedyalkov, N.N.; Stoyanchov, T.R.; Kolev, K.N.; Valova, E.I.; Georgieva, J.S.; Armyanov, S.A.; Amoruso, S.; Wang, X.; et al. fs- and ns-laser processing of polydimethylsiloxane (PDMS) elastomer: Comparative study. *Appl. Surf. Sci.* **2015**, *336*, 321–328. [CrossRef]
20. Lipperta, T.; Wei, J.; Wokaun, A.; Hoogen, N.; Nuyken, O. Polymers designed for laser microstructuring. *Appl. Surf. Sci.* **2000**, *168*, 270–272. [CrossRef]
21. Ahmed, N.; Darwish, S.; Alahmari, A.M. Laser ablation and laser-hybrid ablation processes: A review. *Mater. Manuf. Process.* **2016**, *31*, 1121–1142. [CrossRef]
22. Lippert, T.; Yabe, A.; Wokaun, A. Laser ablation of doped polymer systems. *Adv. Mater.* **1997**, *9*, 105–119. [CrossRef]
23. Shen, H.; Liu, J.; Chen, Y.; Zhang, J.; Zhang, Z.; Guan, N.; Zhang, F.; Huang, L.; Zhao, D.; Jin, Z.; et al. Investigation on time stability of laser-textured patterned surfaces under different temperatures. *Surf. Coat. Technol.* **2020**, *400*, 126225. [CrossRef]
24. Zhou, J.; Xu, R.; Jiao, H.; Bao, J.; Liu, Q.; Long, Y. Study on the mechanism of ultrasonic-assisted water confined laser micromachining of silicon. *Opt. Lasers Eng.* **2020**, *132*, 106118. [CrossRef]
25. Bityurin, N.; Luk'yanchuk, B.S.; Hong, M.H.; Chong, T.C. Models for laser ablation of polymers. *Chem. Rev.* **2003**, *103*, 519–552. [CrossRef]
26. Kim, W.G.; Kim, D.W.; Tcho, I.W.; Kim, J.K.; Kim, M.S.; Choi, Y.K. Triboelectric nanogenerator: Structure, mechanism, and applications. *ACS Nano* **2021**, *15*, 258–287. [CrossRef]
27. Burgess, S.K.; Leisen, J.E.; Kraftschik, B.E.; Mubarak, C.R.; Kriege, R.M.; Koros, W.J. Chain mobility, thermal, and mechanical properties of poly(ethylene furanoate) compared to poly(ethylene terephthalate). *Macromolecules* **2014**, *47*, 1383–1391. [CrossRef]
28. Shin, B.; Oh, J.Y. Photothermal and photochemical investigation on laser ablation of the polyimide by 355nm UV laser processing. *J. Korean Soc. Precis. Eng.* **2007**, *24*, 147–152.
29. Parvinzadeh, M.; Moradian, S.; Rashidi, A.; Yazdanshenas, M.E. Surface characterization of polyethylene terephthalate/silica nanocomposites. *Appl. Surf. Sci.* **2010**, *256*, 2792–2802. [CrossRef]
30. Mohan, J. *Organic Spectroscopy: Principle & Application*; Mehra: New Delhi, India, 2000.
31. Bistričić, L.; Borjanović, V.; Leskovac, M.; Mikac, L.; McGuire, G.E.; Shenderova, O.; Nunn, N. Raman spectra, thermal and mechanical properties of poly(ethylene terephthalate) carbon-based nanocomposite films. *J. Polym. Res.* **2015**, *22*, 39. [CrossRef]
32. Sattmann, R.; Mönch, I.; Krause, H.; Noll, R.; Couris, S.; Hatziapostolou, A.; Mavromanolakis, A.; Fotakis, C.; Larrauri, E.; Miguel, R. Laser-Induced Breakdown Spectroscopy for Polymer Identification. *Appl. Spectrosc.* **1998**, *52*, 456–461. [CrossRef]
33. Mansour, N.; Ghaleh, K.J. Ablation of polyethylene terephthalate at 266 nm. *Appl. Phys. A* **2002**, *74*, 63–67. [CrossRef]
34. Watanabe, H.; Yamamoto, M. Laser ablation of poly(ethylene terephthalate). *J. Appl. Polym. Sci.* **1997**, *64*, 1203–1209. [CrossRef]
35. Klose, S.; Arenholz, E.; Heitz, J.; Bäuerle, D. Laser-induced dendritic structures on PET (polyethylene- terephthalate): The importance of redeposited ablation products. *Appl. Phys. A* **1999**, *69*, 487–490. [CrossRef]
36. Wu, C.; Xu, J.; Zhang, T.; Xin, G.; Li, M.; Rong, Y.; Zhang, G.; Huang, Y. Precision cutting of PDMS film with UV-nanosecond laser based on heat generation-diffusion regulation. *Opt. Laser Technol.* **2022**, *145*, 107462. [CrossRef]

micromachines

MDPI

Article

Modulation and Control of Wettability and Hardness of Zr-Based Metallic Glass via Facile Laser Surface Texturing

Qinghua Wang [1,2], Yangyang Cheng [3], Zhixian Zhu [1,2], Nan Xiang [1,2] and Huixin Wang [4,*]

[1] School of Mechanical Engineering, Southeast University, Nanjing 211189, China; qinghua-wang@seu.edu.cn (Q.W.); zxzhu@seu.edu.cn (Z.Z.); nan.xiang@seu.edu.cn (N.X.)
[2] Jiangsu Key Laboratory for Design and Manufacture of Micro-Nano Biomedical Instruments, Southeast University, Nanjing 211189, China
[3] Guangdong University of Science and Technology Coordination and Innovation Research Institute, Foshan 528000, China; ggdcnc@163.com
[4] Institute of Agricultural Facilities and Equipment, Jiangsu Academy of Agricultural Sciences, Nanjing 210014, China
* Correspondence: wanghx17@mails.jlu.edu.cn; Tel.: +86-1-894-033-1946

Abstract: Bulk metallic glass (BMG) has received consistent attention from the research community owing to its superior physical and mechanical properties. Modulating and controlling the surface functionalities of BMG can be more interesting for the surface engineering community and will render more practical applications. In this work, a facile laser-based surface texturing technique is presented to modulate and control the surface functionalities (i.e., wettability and hardness) of Zr-based BMG. Laser surface texturing was first utilized to create periodic surface structures, and heat treatment was subsequently employed to control the surface chemistry. The experimental results indicate that the laser textured BMG surface became superhydrophilic immediately upon laser texturing, and it turned superhydrophobic after heat treatment. Through surface morphology and chemistry analyses, it was confirmed that the wettability transition could be ascribed to the combined effects of laser-induced periodic surface structure and controllable surface chemistry. In the meantime, the microhardness of the BMG surface has been remarkably increased as a result of laser surface texturing. The facile laser-based technique developed in this work has shown its effectiveness in modification and control of the surface functionalities for BMG, and it is expected to endow more useful applications.

Keywords: laser surface texturing; wettability; hardness; Zr-based metallic glass

Citation: Wang, Q.; Cheng, Y.; Zhu, Z.; Xiang, N.; Wang, H. Modulation and Control of Wettability and Hardness of Zr-Based Metallic Glass via Facile Laser Surface Texturing. *Micromachines* **2021**, *12*, 1322. https://doi.org/10.3390/mi12111322

Academic Editors: Youmin Rong, Congyi Wu and Yu Huang

Received: 27 September 2021
Accepted: 27 October 2021
Published: 28 October 2021

1. Introduction

Bulk metallic glass (BMG) has received considerable attention from the research community during the past several decades since its first discovery in the 1990s, mainly owing to its superior mechanical and physical properties [1], including high values of yield strength [2], high hardness [3], relatively low Young's modulus [4], good corrosion and wear resistance [5], as well as excellent magnetic properties [6]. This gives BMG a variety of potential applications in the fields of bioimplants, magnetic materials, structural materials, sensors, microelectromechanical systems (MEMS), and micro/macro devices [7,8].

Besides its intrinsic outstanding physical and mechanical properties, researchers have also attempted to modify the surface functionalities of BMG, which can be achieved by introducing micro/nanostructures into BMG. The typical fabrication methods of surface structuring include magnetron sputtering [9,10], electro-oxidation [11], thermoplastic shaping [12–14], and laser surface texturing [15,16]. Among all the existing surface modification methods, laser surface texturing has demonstrated its strong potential as a highly efficient and cost-effective approach due to several key advantages including process efficiency, flexibility, ease for automation, and environmental friendliness [17].

In recent years, laser-based surface texturing has been proved to be one of the most efficient techniques to modify and control important surface functionalities, including surface wettability [18–22], reflectivity [23,24], anti-icing property [25,26], corrosion resistance [27], etc. For example, wettability transition from superhydrophilicity to superhydrophobicity has been achieved on various materials including aluminum [28], copper [29], stainless steel [30,31], and titanium [32–36] by combining laser surface texturing and low-temperature annealing. In terms of laser texturing for BMG materials, there have been some recent research efforts on modification of surface properties of BMG via creation of laser-induced surface texture and change of surface chemistry [15,16,37–42]. Huang et al. utilized nanosecond pulsed laser irradiation to fabricate hierarchical micro/nanostructures on the Zr-based metallic BMG substrate in order to increase the effective surface area [16]. The laser-modified BMG surface retained amorphous characteristics, and the elemental distribution on the surface was very uniform. Jiao et al. developed a nanosecond laser texturing technique to fabricate periodic surface structures, including dimples and grooves on Zr-based BMG surfaces [37]. They also investigated the effect of laser surface texturing on the wettability [15] and cytocompatibility [38] of the BMG surfaces. The modification of the surface wettability could be attributed to the laser-induced surface roughness and alteration of surface chemistry, and the enhanced cytocompatibility of the groove-textured BMG resulted from the combined effects of surface chemistry, wettability, and roughness. Du et al. fabricated laser-induced periodic surface structure (LIPSS) and nanoparticle structures on four types of Zr-based BMGs using femtosecond laser irradiation [40]. The experimental results indicated that the femtosecond laser nanostructured Zr-based BMG surface could lead to a distinct decrease in bacterial adhesion compared with the polished surfaces, which was strongly related to the laser-induced surface morphology and wettability. Although the above-mentioned research efforts have effectively modified the surface properties of BMG and achieved improved surface functionalities, none of them has attempted to realize the precise control of the key surface functions of BMG, e.g., control of surface wettability and microhardness, which could be more attractive and challenging, and also help to meet more different applications [43]. Further exploration of a time-efficient and cost-effective laser-based technique to realize the control of surface functionalities on the BMG surface is still of particular interest for the surface engineering community.

Previously, the authors' group has developed a facile nanosecond laser-based surface texturing method to achieve switchable wettability control of titanium alloy [36]. In this work, this laser-based surface texturing technique was further extended to modulate surface wettability and hardness on the BMG substrate. The periodic surface textures were directly created via laser texturing, and the surface chemistry was effectively controlled via heat treatment. The surface wettability was shown to convert from superhydrophilicity immediately upon lase texturing to superhydrophobicity after heat treatment, and the surface microhardness was significantly enhanced on the laser-induced surface texture. The underlying processing mechanisms were elucidated using scanning electron microscopy (SEM), energy-dispersive X-ray spectroscopy (EDAX), and X-ray photoelectron spectroscopy (XPS). Compared with the existing research works on laser texturing for functional surfaces, the laser-based surface texturing technique developed in this work proposed a novel and highly efficient approach to modulate and control the surface functionalities, which was achieved by the combination of high-speed UV nanosecond laser surface texturing and subsequent heat treatment. It is expected that the developed technique could provide a viable solution for the surface modification of Zr-based metallic glass, thus rendering a series of applications in the industrial and biomedical fields.

2. Materials and Methods

2.1. Materials

Commercially available Zr-based bulk metallic glass Vitreloy 1 with the nominal elemental composition $Zr_{41.2}Ti_{13.8}Cu_{12.5}Ni_{10}Be_{22.5}$ (at%, purchased from METALLAB, Changzhou, China) was used in this work due to its excellent mechanical properties. The samples were mechanically grinded using grid SiC papers and further machined into thin sheets with a dimension of 30 mm × 30 mm × 2 mm using wire electrical discharge machining. Then, before laser surface texturing experiments, they were ultrasonically cleaned with acetone, ethanol, and deionized water successively to remove the contaminants.

2.2. Laser-Based Surface Texturing

The laser-based surface texturing experiment mainly consists of two steps: (1) laser surface texturing and (2) heat treatment. Laser surface texturing experiments employed a laser marking machine (TH-UV200A, Tianhong Laser, Suzhou, China) equipped with a 355 nm UV laser source (AWAVE 355-15W-30K, Advanced Optowave, Ronkonkoma, NY, USA). The laser source emits a laser beam guided by reflective mirrors. The intensity of laser power is controlled by the attenuator, and the diameter of the laser beam can be expanded by the beam expander. The focusing lens of the system provides a focal spot diameter of ~35 μm during the laser texturing experiments. Unidirectional line pattern and cross-hatch pattern were used due to the ease and high efficiency of fabrication, as shown in Figure 1a,b. A series of preliminary experimental trials was attempted, and the optimal laser processing parameters were determined and utilized in this work, which can be found in Table 1. The schematics for the laser surface texturing experiments are shown in Figure 2a.

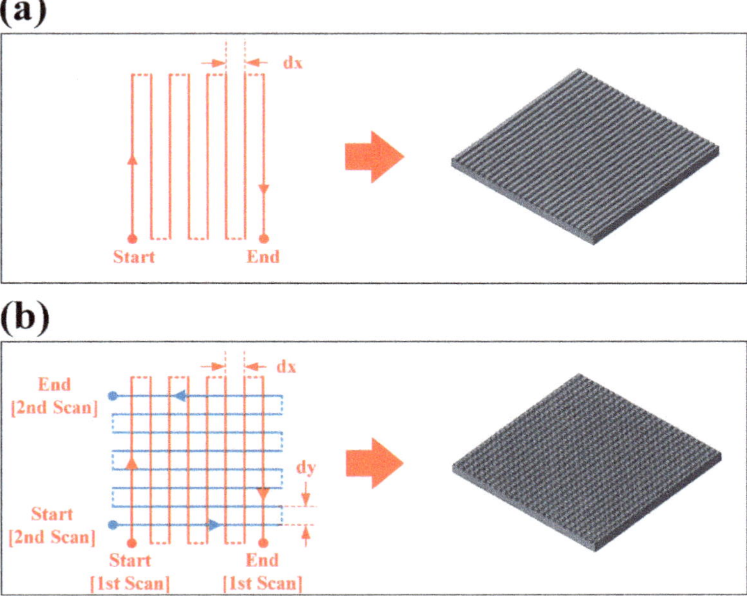

Figure 1. Pattern designs used for laser surface texturing of BMG: (**a**) unidirectional; (**b**) cross-hatch.

Table 1. Processing parameters for the laser-based surface texturing experiments.

Name of Parameter	Value
Average power (W)	9.0, 10.5, 12.0, 13.5, 15.0
Repetition rate (kHz)	30
Pulse width (ns)	20
Scanning speed (mm/s)	20, 40, 60, 80
Step size (μm)	150
Power intensity (GW/cm^2)	1.56~2.60
Pulse energy (mJ)	0.3~0.5
Heat treatment temperature (°C)	150
Heat treatment duration (h)	2

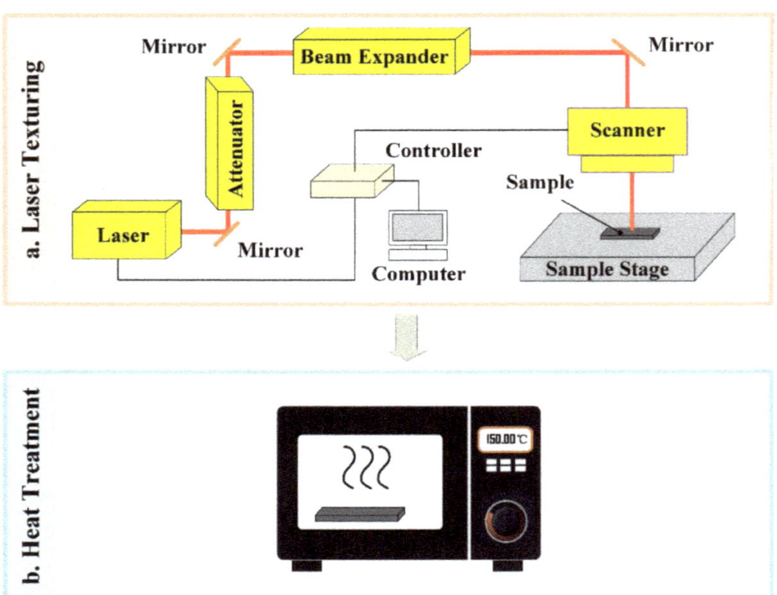

Figure 2. Schematics for the laser-based surface texturing experiments of BMG: (**a**) laser surface texturing; (**b**) heat treatment.

Immediately upon laser surface texturing, the laser textured BMG surfaces became superhydrophilic. To achieve the wettability transition, the laser textured BMG surfaces were placed in a conventional furnace for low-temperature annealing treatment, as shown in Figure 2b. An annealing temperature of 150 °C and a treatment duration of a maximum of 2 h were used during the heat treatment.

2.3. Surface Characterizations

The surface morphology of the laser textured BMG surfaces was analyzed using field emission scanning electron microscopy (FESEM, Navo Nano SEM450, Hillsboro, OR, USA). The surface chemistry of the laser textured surface was evaluated using energy dispersive X-ray analysis (EDAX, FEI Sirion, Hillsboro, OR, USA) and X-ray photoelectron spectroscopy (XPS, PREVAC, Rogów, Poland). To examine the surface wettability of the laser textured surfaces, a contact angle goniometer (SDC-200, Sindin Precision Instrument Co., Ltd., Dongguan, Guangdong, China) equipped with a high-resolution CMOS camera was utilized to measure the static water contact angle value (θ_W) on each surface. A water droplet with the volume of ~4 µL was dripped onto the sample surface using a digital syringe during each measurement. The optical image of the water contact angle measurement was captured using the CMOS camera, and image analysis software was utilized to determine the θ_W value for each measurement. For each sample, four θ_W measurements were performed at various locations, and the averaged θ_W value was reported. The standard deviation for each averaged θ_W measurement result was also calculated and added. The microhardness test was carried out using a Vickers hardness machine (HXD-1000TMSC/LCD, Shanghai, China) with a test load of 300 gf and a dwell time of 10 s. Similarly, five microhardness measurements were performed at different locations on the laser textured surface, and the averaged hardness value as well as the standard deviation for each measurement result were reported.

3. Results

3.1. Surface Morphology

The surface morphology of the laser textured BMG surface using different scanning speeds was examined by SEM, as shown in Figure 3. The SEM micrographs of the laser textured BMG surfaces with unidirectional surface patterns and various scanning speeds can be found in Figure 3a–d, while the SEM micrographs of the laser textured BMG surfaces with cross-hatch surface patterns and various scanning speeds are shown in Figure 3e–h. It is clearly seen that as the laser beam irradiated and ablated the BMG substrates, periodic micro-scale bulge structures were formulated as a result of the strong interaction between the laser beam and the BMG substrates. The SEM micrographs with high magnification indicate that there also have been some sub-micron or nano-scale particles covered on the micro-bulge structures, which can be mainly attributed to the ejection and deposition of nanoparticles during the laser-material interaction. In addition, it can be found that the laser-induced periodic surface structure exhibited distinct differences as the scanning speed increased. When the scanning speed of 20 mm/s was used, solid and defect-free micro-bulge structures were formed on the laser textured BMG surfaces with both unidirectional and cross-hatch patterns. As the scanning speed further increased, the surface structure gradually changed, which was illustrated by the variation of the micro-bulge height and the appearance of concave sections. Especially when the scanning speeds reached 60 mm/s and 80 mm/s, the concave structure became more distinct along the laser scanned line profiles for both surface patterns. The difference of the surface structure formation when varying the scanning speed should result from the time duration of laser–material interaction. When the lower scanning speed was utilized, the laser beam travelled relatively slowly along the scanned profile, and the number of pulses per irradiation point is higher, which would ensure adequate interaction between the laser beam and the BMG substrate and facilitate the formation of solid micro-bulge structures [44]. However, as the scanning speed increased, the laser beam moved faster and resulted in the decrease in the number of pulses per irradiation point. This would significantly weaken the impact of the laser beam on the BMG substrate. Consequently, due to the insufficient interaction time between the laser beam and the BMG substrate, fewer sub-micron or nano-scale particles and concave structures occurred on the laser textured surface with higher scanning speeds.

The effect of laser power on the surface morphology of the laser textured BMG surface was further examined by SEM, as shown in Figure 4. The SEM micrographs of the laser textured BMG surfaces with unidirectional surface patterns and various laser powers can be found in Figure 4a–e, while the SEM micrographs of the laser textured BMG surfaces with cross-hatch surface patterns and various laser powers are shown in Figure 4e–h. From the SEM micrographs, it is evident that laser power can dramatically affect the formation of laser-induced structures on the BMG surface. By using higher laser powers (15.0 W and 13.5 W), clear concave microgrooves formed on the BMG surface, demonstrating strong vaporization of the material during the laser surface texturing process. As the laser power decreased (12.0 W and 10.5 W), the microgroove structure became less pronounced, only with some holes left on the laser scanned profiles. This indicated that less material evaporated from the substrate, and there appeared to be a restructuring on the laser textured surface. As the laser power kept decreasing (9.0 W), the concave structure almost disappeared, and the convex micro-bulge structure dominated. Similar trends could be observed for the laser textured BMG surfaces with both unidirectional and cross-hatch surface patterns. As clearly pointed out by Feng et al. [45,46], the laser-induced surface structure is a key factor that will affect the surface functionality, i.e., wettability, reflectivity, and hardness, and the laser processing parameters must be carefully chosen to ensure the formation of proper laser-induced structure, thus enabling the realization of desirable surface functionalities.

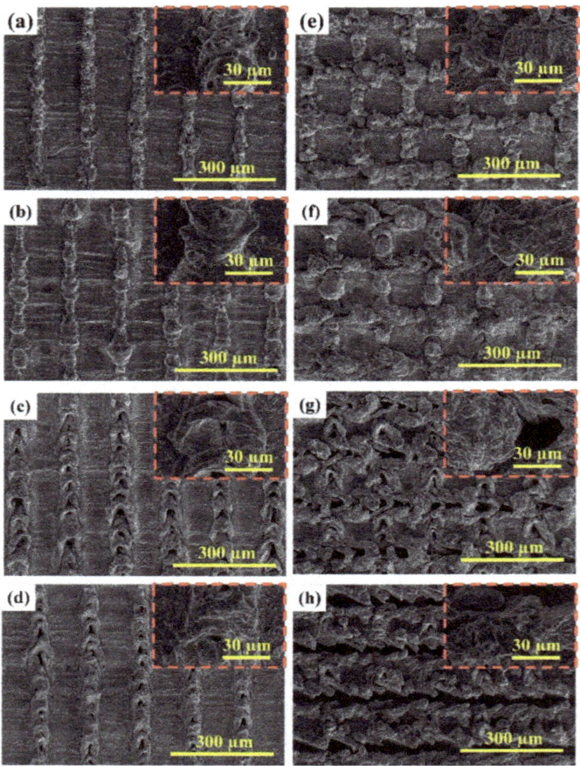

Figure 3. SEM micrographs of the laser textured BMG surfaces using different scanning speeds: (**a–d**) unidirectional surface patterns with the scanning speeds of 20 mm/s, 40 mm/s, 60 mm/s, and 80 mm/s; (**e–h**) cross-hatch surface patterns with the scanning speeds of 20 mm/s, 40 mm/s, 60 mm/s, and 80 mm/s.

3.2. Surface Wettability

Surface wettability of the BMG surfaces with different surface patterns and treatment methods were evaluated via contact angle measurements. It can be found that the untreated BMG surface exhibited a θ_w of 87.4 ± 1.8°, as shown in Figure 5a, indicating that the untreated BMG surface is hydrophilic. Immediately upon laser surface texturing using the processing parameters of a laser power of 12 W, a repetition rate of 30 kHz, and a scanning speed of 40 mm/s, the laser textured BMG surfaces with both unidirectional (Figure 5b) and cross-hatch (Figure 5c) surface patterns exhibited a θ_w of 0°. This clearly indicates that the laser textured BMG surfaces are in a saturated Wenzel regime when being treated in the oxygen-containing atmosphere, which agrees well with the experimental results in [47]. After heat treatment for 1 h, the θ_w measurements indicated that the laser textured BMG surfaces with unidirectional and cross-hatch patterns exhibited θ_w values of 145.5 ± 1.9° and 145.7 ± 2.5°, respectively, indicating that a 1 h heat treatment rendered the laser textured BMG with high hydrophobicity. With a 2 h heat treatment, the laser textured BMG surfaces with both unidirectional and cross-hatch patterns became superhydrophobic, with θ_w values of 154.3 ± 1.9° and 153.7 ± 1.1°, respectively.

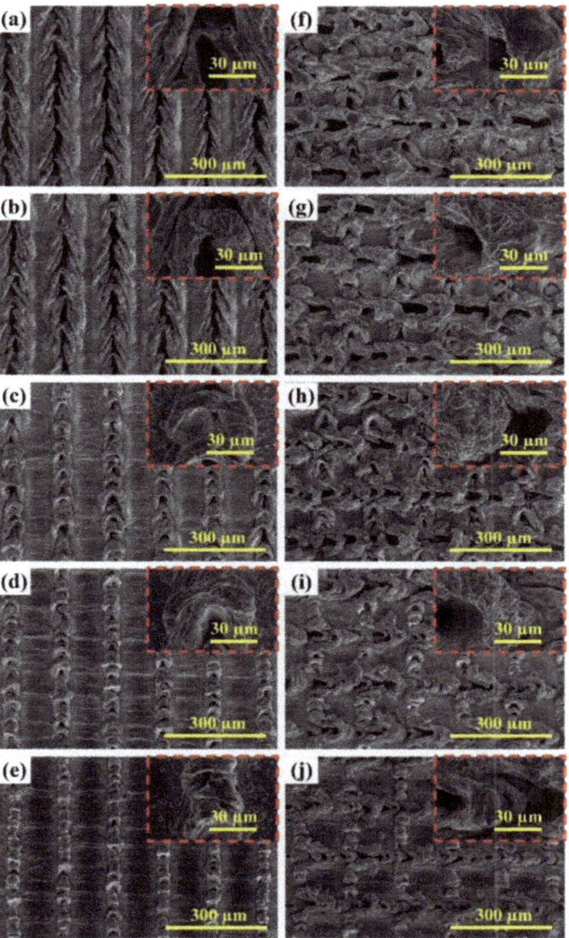

Figure 4. SEM micrographs of the laser textured BMG surfaces using different powers: (**a**–**e**) unidirectional surface patterns with the laser powers of 15.0 W, 13.5 W, 12.0 W, 10.5 W, and 9.0 W; (**f**–**j**) cross-hatch surface patterns with the laser powers of 15.0 W, 13.5 W, 12.0 W, 10.5 W, and 9.0 W.

Figure 6 shows θ_W measurement results for the laser textured BMG with varied laser powers and scanning speeds. As shown in Figure 6a, the untreated BMG surface showed a hydrophilic nature with a θ_W of 84.4 ± 1.2°. For the laser texturing experiments, the following laser processing parameters were utilized: a laser power of 10.5 W, a repetition rate of 30 kHz, and a scanning speed of 60 mm/s. The θ_W measurement results (Figure 6b,c) confirmed the saturated Wenzel regime for the laser textured surfaces with both of the unidirectional and cross-hatch patterns. Subsequently, it can be found that a 1 h heat treatment turned the laser textured surfaces with both patterns highly hydrophobic, with θ_W values of 144.7 ± 1.1° and 146.5 ± 1.7°, respectively (Figure 6d,e). A 2 h heat treatment ensured that both of the laser textured surfaces reached superhydrophobicity, with θ_W values of 155.9 ± 1.2° and 154.5 ± 1.9°, respectively (Figure 6f,g). As discussed in the previous section, low-temperature heat treatment with an annealing temperature of ~150 °C has been proved to be an efficient approach to achieve wettability transition from superhydrophilicity to superhydrophobicity for various metallic materials such as aluminum, copper, stainless steel, and titanium [28–36]. However, this wetting state transition approach has never been utilized and confirmed for BMG. To the authors' best knowledge, this work represents the first attempt to achieve wettability transition from superhydrophilicity to superhydrophobicity on BMG surfaces using the facile laser surface texturing technique combined with low-temperature annealing. The underlying processing mechanisms associated with the wettability will be explained in detail in the following sections.

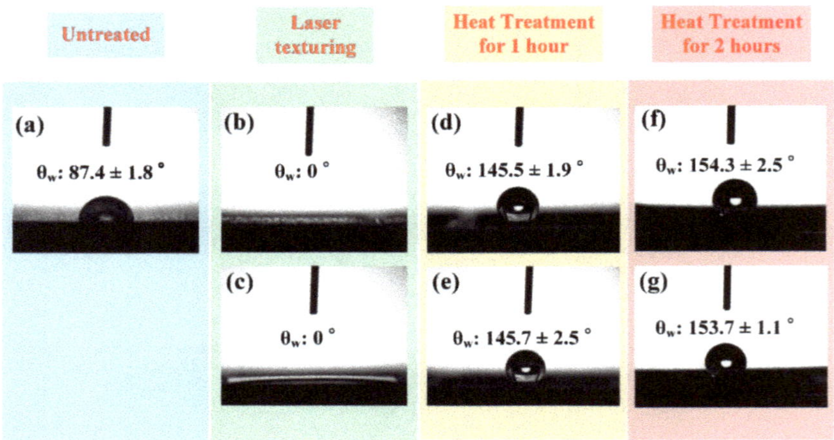

Figure 5. Contact angle measurement results for (**a**) untreated BMG surface; (**b**) laser textured BMG surface using the power of 12 W, the repetition rate of 30 kHz, the scanning speed of 40 mm/s, and a unidirectional surface pattern; (**c**) laser textured BMG surface using the power of 12 W, the repetition rate of 30 kHz, the scanning speed of 40 mm/s, and a cross-hatch surface pattern; (**d**,**e**) laser textured BMG surfaces in (**b**,**c**) after heat treatment for 1 h; (**f**,**g**) laser textured BMG surfaces in (**b**,**c**) after heat treatment for 2 h.

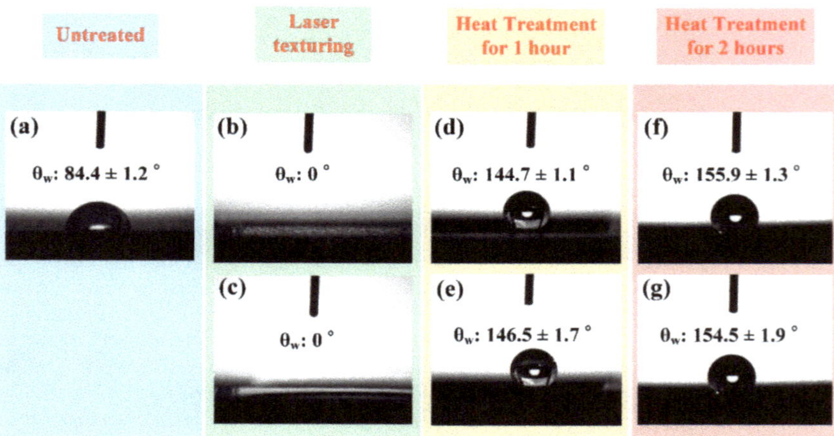

Figure 6. Contact angle measurement results for (**a**) untreated BMG surface; (**b**) laser textured BMG surface using the power of 10.5 W, the repetition rate of 30 kHz, the scanning speed of 60 mm/s, and a unidirectional surface pattern; (**c**) laser textured BMG surface using the power of 12 W, the repetition rate of 30 kHz, the scanning speed of 40 mm/s, and a cross-hatch surface pattern; (**d**,**e**) laser textured BMG surfaces in (**b**,**c**) after heat treatment for 1 h; (**f**,**g**) laser textured BMG surfaces in (**b**,**c**) after heat treatment for 2 h.

3.3. Surface Chemistry

Farasi et al. [48] pointed out that the surface chemical composition of metallic materials can be modified via material vaporization and oxidation during laser surface texturing, and Ngo et al. [31,49] believed that the increase in carbon content during heat treatment indicates the deposition of more hydrophobic functional groups on the laser textured surface, thus leading to the transition of surface wettability. Therefore, to explore and understand the variations of chemical compositions on the BMG surface and to further correlate surface wettability with surface chemistry, the laser textured BMG surfaces before and after heat treatment were analyzed by EDAX, as shown in Figure 7. The laser textured surfaces (with and without heat treatment) considered for the surface chemistry analysis were treated with the average laser power of 12.0 W, repetition rate of 30 kHz, pulse energy of 0.4 mJ, and power intensity of 2.08 GW/cm^2. From Figure 7, it can be clearly seen that all the core elements can be detected on the three different types of surfaces. While compared with the untreated surface, the intensity of the Zr peak for the laser textured surface decreased, which is mainly a result of material vaporization during laser texturing. In the meantime, the O peak shows a notable increase in the laser textured surface compared with that of the untreated surface, indicating the BMG surface has been oxidized along with the formation of a periodic surface structure. As indicated in [50], as the electronic structure of metal oxide facilitates the formation of hydrogen bonds and increases the surface energy, the laser textured surface typically exhibited superhydrophobicity. For the laser textured surface with heat treatment, subtle increases in C and Si peaks can be detected, indicating that hydrophobic functional groups, including $-CH_2-$, $-CH_3$, C=C as well as thin PDMS layer, should have been absorbed and deposited onto the laser textured BMG surface, rendering the heat-treated surface with superhydrophobicity [33,49].

The EDAX mapping data were also obtained to reveal the elemental distribution of all the core elements (Zr, Ti, Be, Cu, Ni, Si, C, and O) on each surface, as shown in Figure 8. From EDAX mapping, it can be clearly observed that all the core elements were uniformly distributed on the untreated surface. For the laser textured surfaces, more black areas can be observed, suggesting a reduction of the Zr, Ti, Be, Cu, and Ni elements, corresponding to the material removal during laser surface texturing. Distinct increases for the amount of O can be seen on the EDAX mapping as well, which strongly supports the oxidation during laser surface texturing. For the elemental distribution of C and Si, no clear change can be observed, which will be further examined using XPS.

Given the fact that EDAX is mainly a qualitative analysis method that could not provide conclusive results for the evaluation of the BMG substrate after laser texturing and heat treatment, XPS analysis was further employed to investigate the chemical changes on the surfaces that have been tested by EDAX. The XPS full spectra of the BMG surfaces considered for EDAX analysis and the corresponding atomic percentage of different elements are shown in Figure 9. It can be seen from the XPS full spectra of the untreated BMG surface (Figure 9a) that carbon and oxygen were detected as the major elements, which can be attributed to the contamination and oxidation of the BMG surface. For the constituent elements of the BMG Vitreloy 1 used in this work, small peaks of Zr 3d and Be 1s were observed on the untreated BMG surface, while the content of other elements (Ti 2p, Ni 2p, and Cu 2p) was negligible. Upon laser texturing, the atomic percentage of oxygen increased from 26.88% to 47.85% (Figure 9b). This clearly indicates that the laser textured BMG surface has been oxidized along with the formation of periodic surface structure during laser texturing. As the hydrogen bonds tend to form on the metal oxide, which also increases the surface energy [50], the laser oxidized BMG surface exhibited superhydrophilicity. After heat treatment, a clear atomic percentage increase in carbon from 30.85% to 37.80% can be observed on the laser textured BMG surface with heat treatment (Figure 9c). The increase in carbon content demonstrated that several different types of functional groups that exhibit hydrophobicity could have been absorbed onto the laser textured BMG surface during heat treatment, which rendered the heat-treated BMG surface highly hydrophobic [33,49]. In addition, the element Si was also detected on the heat-treated BMG surface, as shown in

Figure 9c. The appearance of Si might represent the formation of a thin Si-based PDMS layer on the laser textured BMG surface during heat-treatment, which further increased the hydrophobicity of the laser textured BMG surface. The experimental findings in this work agree well with the results of Ti alloy using a similar method [36], and it is believed that the increase in carbon and the appearance of a thin PDMS layer contribute to the superhydrophobicity of the laser textured BMG surface after heat treatment.

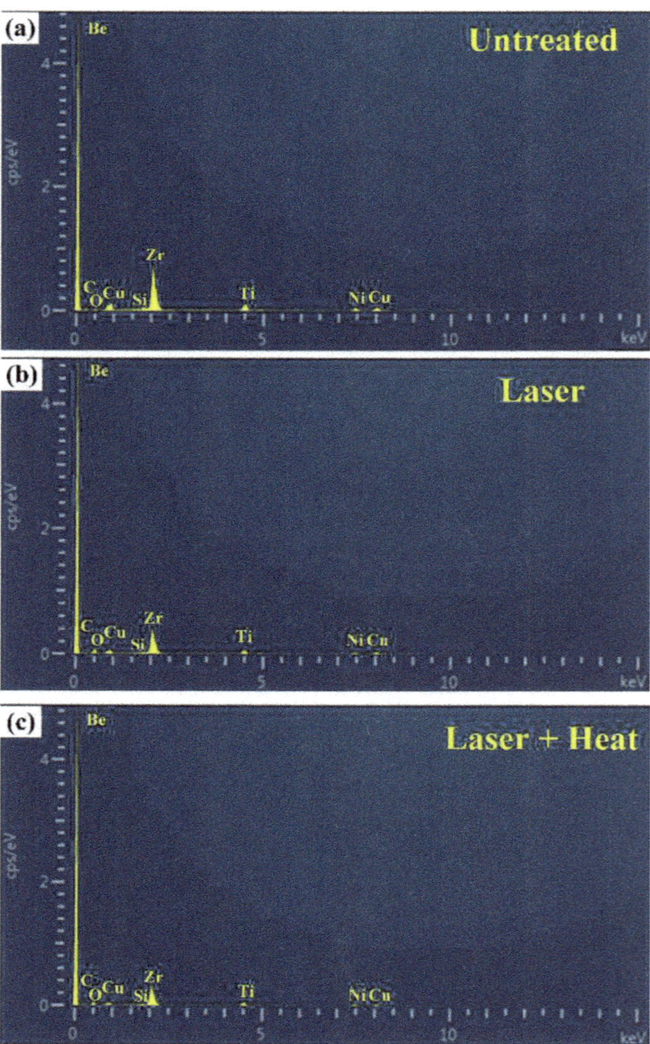

Figure 7. EDAX spectra for (**a**) untreated BMG surface; (**b**) laser textured BMG surface; (**c**) laser textured BMG surface followed by heat treatment.

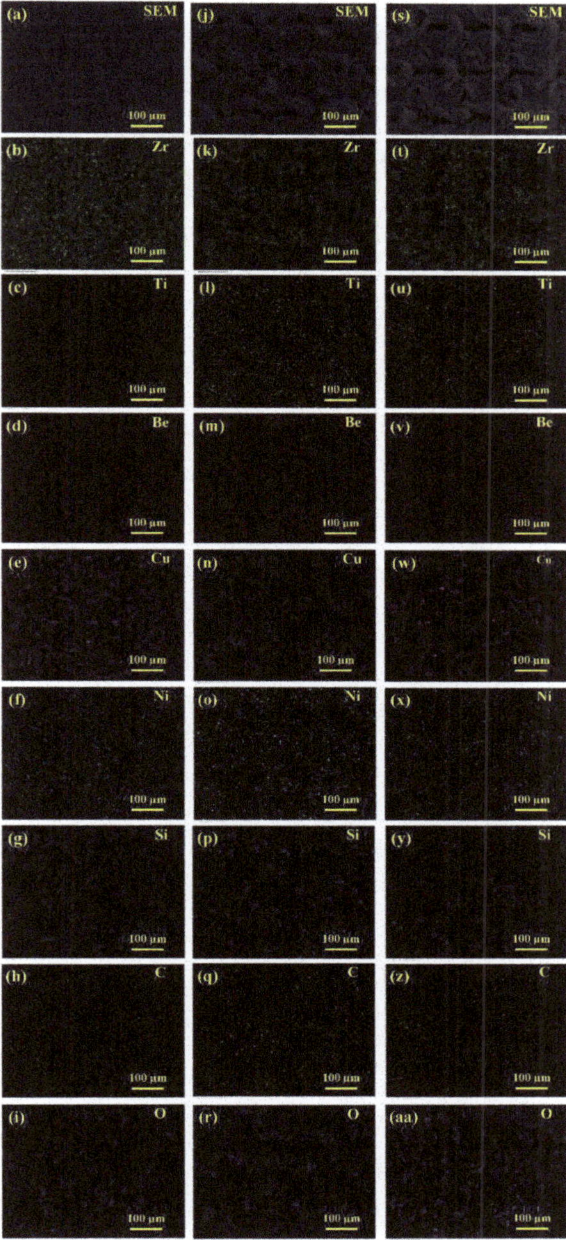

Figure 8. SEM/EDAX element mapping for (**a**–**i**) untreated BMG surface; (**j**–**r**) laser textured BMG surface; (**s**–**aa**) laser textured BMG surface followed by heat treatment. (**a**,**j**,**s**) are SEM micrographs representing the corresponding analyzed regions; (**b**–**i**,**k**–**r**,**t**–**aa**) are the EDAX maps showing the qualitative elemental distributions of Zr, Ti, Be, Cu, Ni, Si, C, and O.

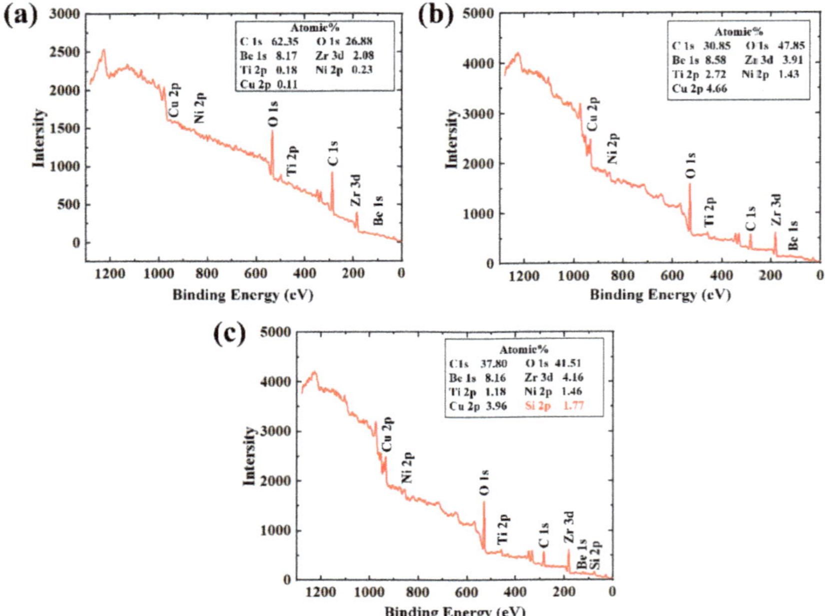

Figure 9. XPS full spectra of (**a**) untreated BMG surface; (**b**) laser textured BMG surface; (**c**) laser textured BMG surface followed by heat treatment.

Core elemental analyses were also performed for C, O, and Si elements, as shown in Figure 10. The core elemental analysis for C element can be found in Figure 10a–c. In Figure 10a, it is clearly shown that C element can be detected, confirming the contamination of the untreated BMG surface. For the laser textured BMG surfaces, it can be observed in Figure 10b and c that after heat treatment, the proportion of C–C and C–H peaks significantly increased, which helped to verify that hydrophobic functional groups with nonpolar C–C or C–H bonds should have been deposited onto the heat-treated BMG surface, inducing hydrophobicity on the surface. The core elemental analysis for O is shown in Figure 10d–f. It is clear that functional groups of $(OH)^-$ and O^{2-} were detected on all of the BMG surface, while the proportion of O^{2-} on the laser textured BMG surfaces both with and without heat treatment was higher than that of the untreated BMG surface, proving that oxidation occurred during the laser texturing process. Finally, Figure 10g shows the core elemental analysis for Si element, revealing the appearance of the peaks for three functional groups Si–O, Si–C, and –Si– on the laser textured BMG surface after heat treatment. This should be attributed to the thin PDMS layer deposited on the laser textured BMG surface during heat treatment, which is derived from the silicone component on the furnace used in this work, as indicated in [32]. The underlying mechanism will be explained with details in the following section.

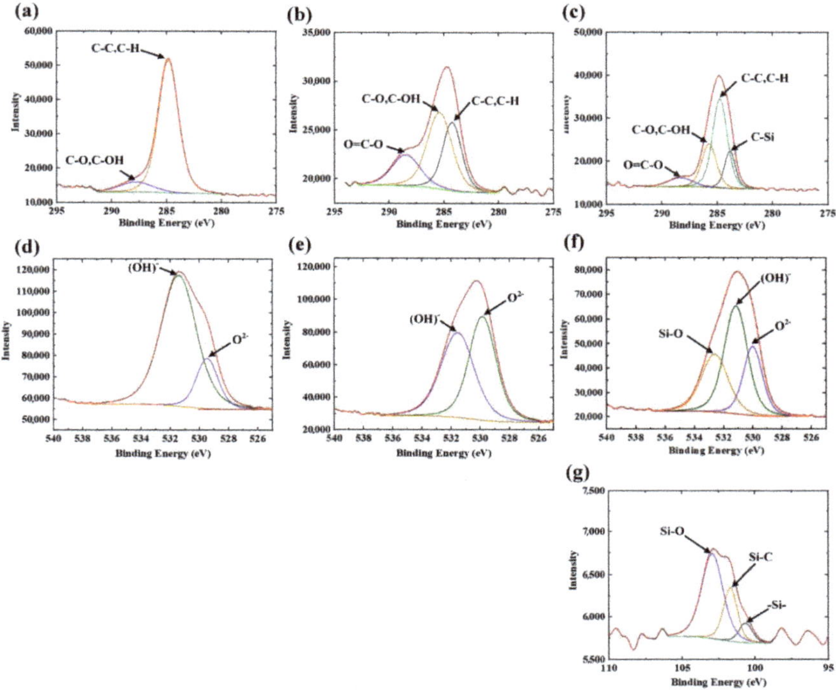

Figure 10. The high-resolution core elemental spectra of carbon on (**a**) untreated BMG surface; (**b**) laser textured BMG surface; (**c**) laser textured BMG surface followed by heat treatment. The high-resolution core elemental spectra of oxygen on (**d**) untreated BMG surface; (**e**) laser textured BMG surface; (**f**) laser textured BMG surface followed by heat treatment; and (**g**) the high-resolution core elemental spectra of silicon on the laser textured BMG surface followed by heat treatment.

3.4. Surface Microhardness

The microhardness measurement results of the untreated BMG surface, the laser textured BMG surface with unidirectional pattern, and the laser textured BMG surface with cross-hatch pattern can be found in Figure 11. The laser textured surfaces considered for the surface microhardness analysis were treated using the same processing parameters as those used for surface chemistry analysis. The experimental results indicated that the microhardness of the untreated BMG surface was 514.4 $\pm$ 4.9 HV. By laser surface texturing using the unidirectional surface pattern, the microhardness of the textured surface was increased to 564.9 $\pm$ 5.2 HV, representing an increase rate of 9.8% compared with the untreated surface. Meanwhile, the microhardness of the laser textured BMG surface with the cross-hatch pattern was further enhanced up to 596.7 $\pm$ 5.8 HV, indicating an increase rate of 16.0% compared with the untreated surface. As evident from the SEM micrographs shown in Figures 3 and 4, the laser surface texturing process created a combination of micro-scale, sub-micron, and nano-scale structures on the BMG substrate, which have generated much finer grains on the surface. As indicated by the Hall–Petch relationship, finer grains will result in higher mechanical strength [51,52]. More grain boundaries can be generated using finer grains, and the resistance to hinder dislocation motion will increase, leading to the increase in surface microhardness. In addition, compared with the laser textured surface with the unidirectional pattern, one more laser scan from the vertical direction would significantly increase the density of surface structures at all scales on the BMG substrate, which would contribute to the further enhancement of microhardness [53]. The microhardness measurement results indicate that the developed laser-based surface texturing process can effectively enhance the surface mechanical strength of the BMG substrate.

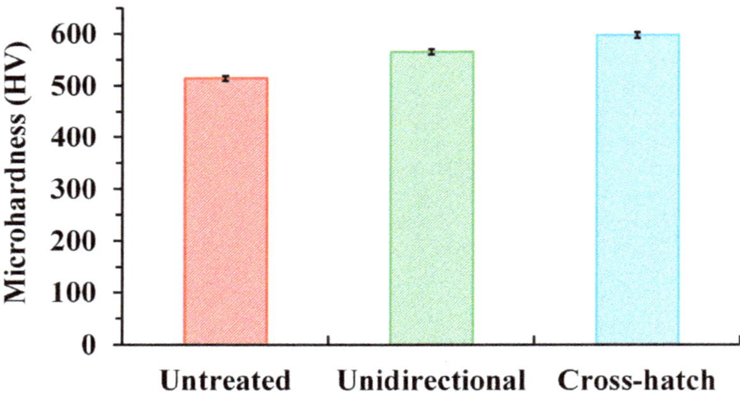

Figure 11. Microhardness measurements for untreated surface, laser textured BMG surface with unidirectional pattern, and laser textured BMG surface with cross-hatch pattern.

3.5. Processing Mechanism Analysis

The underlying mechanism for the wettability transition of BMG surface during laser texturing and heat treatment is schematically depicted in Figure 12. As shown in Figure 12a, nanosecond laser surface texturing not only generated periodic surface structures on the BMG substrate but also oxidized the surface in the ambient air with the existence of water vapor [35,47]. Consequently, formation of oxide and hydroxide layers would occur simultaneously on top of the BMG substrate, which will drastically increase the surface energy and lead to superhydrophilicity [28,54]. During heat treatment, the wettability transition from superhydrophilicity to superhydrophobicity can be ascribed to two aspects: on the one hand, it is proposed that the organic compounds existed in air with nonpolar C–C and C–H bonds that have been absorbed onto the laser-induced micro-bulge or groove structures, and this process has been accelerated as the heat treatment temperature increased. The heat treatment helped to eliminate the hydrophilic –OH functional group from the laser textured BMG surface and created more active sites on the surface for the subsequent absorption of organic compounds [55]. As a result, more hydrophobic functional groups, such as $-CH_2-$ and $-CH_3$, have been attracted onto the laser textured BMG surface, leading to the increase in surface hydrophobicity. On the other hand, the XPS analysis confirmed the existence of Si element on the laser textured BMG surface, indicating that a thin PDMS layer could have been deposited on the surface. The source of the Si element was originated from the silicone seal on the furnace, which was partially vaporized and deposited onto the laser textured BMG surface, as shown in Figure 12c. Since the silicon-based organic polymer PDMS is known to be hydrophobic, it could help to further enhance the hydrophobicity of the laser textured BMG surface, leading to superhydrophobicity eventually. It is thus believed that absorption of hydrophobic airborne organic compounds and generation of the thin PDMS layer both contributed to the surface wettability transition on the laser textured BMG surface during heat treatment. Therefore, from an initial contact angle of ~85° on the untreated BMG surface, the surface wettability was converted to superhydrophilicity (~0°) after laser texturing, and superhydrophobicity (>150°) after heat treatment, as shown in Figure 12b.

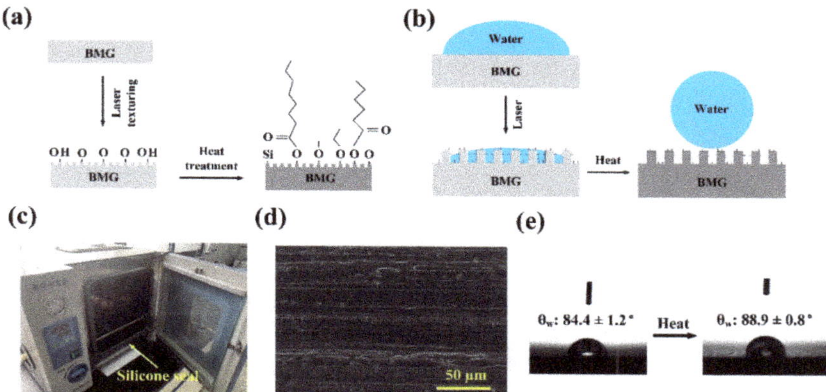

Figure 12. Schematic for the mechanism of wettability transition for the laser textured BMG surface after laser texturing and heat treatment: (**a**) alteration of surface chemistry; (**b**) alteration of surface wettability; (**c**) demonstration of the silicone seal on the furnace used in this work; (**d**) SEM micrograph of the untreated BMG surface; (**e**) contact angle measurement results for the untreated BMG surface before and after heat treatment.

In the meantime, although surface chemistry plays a key role for the wettability transition, it is proposed that the laser-induced surface structure is equally critical for achieving the target wettability condition. This could be verified by what has been observed on the untreated BMG surface before and after heat treatment. Figure 12d shows the SEM micrograph of the untreated BMG surface, indicating that there are only grinding marks without any distinct periodic surface structures. The contact angle measurement results indicated that the initial θ_W was 84.4 $\pm$ 1.2° on the untreated BMG before heat treatment, and the θ_W was measured to be 88.9 $\pm$ 0.8° after heat treatment. This clearly indicates that alteration of surface chemistry alone could not render the BMG material with superhydrophobicity. Surface structure and surface chemistry should be modulated and controlled spontaneously to achieve the desired surface functionality.

4. Conclusions

In this work, a facile and efficient laser-based surface texturing method was developed to modulate and control the surface functionalities of Zr-based BMG. The following findings can be summarized:

(1) The developed laser-based surface texturing technique consists of two steps: laser texturing and heat treatment. Laser texturing generated the periodic surface structures and oxidized the BMG surface, while the subsequent heat treatment accelerated the absorption of hydrophobic airborne organic compounds and deposited a thin PDMS layer on the laser textured BMG surface.

(2) It is found that the untreated BMG surface is hydrophilic with a θ_W of 84.4 $\pm$ 1.2°. Immediately upon laser texturing, the laser textured BMG surface exhibited a θ_W of 0°, indicating the surface turned superhydrophilic. After a heat treatment of 2 h, the laser textured BMG surface became superhydrophobic with a θ_W higher than 150°.

(3) Through careful experimental validation and analysis, it is believed that the laser-induced surface texture and modified surface chemistry are equally important for achieving the desirable wettability condition.

(4) The microhardness of the laser textured BMG surface is also notably increased due to the higher microstructure density and more grain boundaries generated on account of the laser surface texturing process.

This method will provide a feasible and highly efficient solution for regulating and controlling the surface functionalities of BMG, which will render more practical and important applications.

Author Contributions: Conceptualization, Q.W. and H.W.; Data curation, Q.W., Y.C., Z.Z. and H.W.; Formal analysis, Q.W., Y.C. and N.X.; Funding acquisition, Q.W. and H.W.; Investigation, Q.W. and Z.Z.; Methodology, Q.W.; Project administration, Q.W.; Resources, Q.W. and N.X.; Validation, Q.W.; Writing—original draft, Q.W.; Writing—review and editing, H.W. All authors have read and agreed to the published version of the manuscript.

Funding: This research was funded by the Guangdong Basic and Applied Basic Research Foundation (No. 2019A1515110501), the National Natural Science Foundation of China (No. 52105175), the Natural Science Foundation of Jiangsu Province (No. BK20210235) and the New Faculty Start-up Fund of Southeast University (No. 1102007140).

Acknowledgments: The authors would like to acknowledge the Analysis and Testing Center of Southeast University for the help with SEM, EDAX, and XPS analyses.

Conflicts of Interest: The authors declare no conflict of interest.

References

1. Miller, M.; Liaw, P. *Bulk Metallic Glasses*; Springer: Boston, MA, USA, 2008; ISBN 9780387489209.
2. Löffler, J.F. Bulk metallic glasses. *Intermetallics* **2003**, *11*, 529–540. [CrossRef]
3. Telford, M. The case for bulk metallic glass. *Mater. Today* **2004**, *7*, 36–43. [CrossRef]
4. Bian, Z.; Chen, G.L.; He, G.; Hui, X.D. Microstructure and ductile-brittle transition of as-cast Zr-based bulk glass alloys under compressive testing. *Mater. Sci. Eng. A* **2001**, *A316*, 135–144. [CrossRef]
5. Huang, H.; Yan, J. Investigating shear band interaction in metallic glasses by adjacent nanoindentation. *Mater. Sci. Eng. A* **2017**, *A704*, 375–385. [CrossRef]
6. Liu, W.; Zhang, H.; Shi, J.A.; Wang, Z.; Song, C.; Wang, X.; Lu, S.; Zhou, X.; Gu, L.; Louzguine-Luzgin, D.V.; et al. A room-temperature magnetic semiconductor from a ferromagnetic metallic glass. *Nat. Commun.* **2016**, *7*, 13497. [CrossRef] [PubMed]
7. Li, H.F.; Zheng, Y.F. Recent advances in bulk metallic glasses for biomedical applications. *Acta Biomater.* **2016**, *36*, 1–20. [CrossRef] [PubMed]
8. Khan, M.M.; Nemati, A.; Rahman, Z.U.; Shah, U.H.; Asgar, H.; Haider, W. Recent Advancements in Bulk Metallic Glasses and Their Applications: A Review. *Crit. Rev. Solid State Mater. Sci.* **2018**, *43*, 233–268. [CrossRef]
9. Chen, N.; Frank, R.; Asao, N.; Louzguine-Luzgin, D.V.; Sharma, P.; Wang, J.Q.; Xie, G.Q.; Ishikawa, Y.; Hatakeyama, N.; Lin, Y.C.; et al. Formation and properties of Au-based nanograined metallic glasses. *Acta Mater.* **2011**, *59*, 6433–6440. [CrossRef]
10. Ketov, S.V.; Shi, X.; Xie, G.; Kumashiro, R.; Churyumov, A.Y.; Bazlov, A.I.; Chen, N.; Ishikawa, Y.; Asao, N.; Wu, H.; et al. Nanostructured Zr-Pd metallic glass thin film for biochemical applications. *Sci. Rep.* **2015**, *5*, 7799. [CrossRef]
11. Zhao, M.; Abe, K.; Yamaura, S.; Yamamoto, Y.; Asao, N. Fabrication of Pd−Ni−P Metallic Glass Nanoparticles and Their Application as Highly Durable Catalysts in Methanol Electro-oxidation. *Chem. Mater.* **2014**, *26*, 1056–1061. [CrossRef]
12. Kumar, G.; Tang, H.X.; Schroers, J. Nanomoulding with amorphous metals. *Nature* **2009**, *457*, 868–872. [CrossRef] [PubMed]
13. Hu, Z.; Gorumlu, S.; Aksak, B.; Kumar, G. Patterning of metallic glasses using polymer templates. *Scr. Mater.* **2015**, *108*, 15–18. [CrossRef]
14. He, P.; Li, L.; Wang, F.; Dambon, O.; Klocke, F.; Flores, K.M.; Yi, A.Y. Bulk metallic glass mold for high volume fabrication of micro optics. *Microsyst. Technol.* **2016**, *22*, 617–623. [CrossRef]
15. Jiao, Y.; Brousseau, E.; Shen, X.; Wang, X.; Han, Q.; Zhu, H.; Bigot, S.; He, W. Investigations in the fabrication of surface patterns for wettability modification on a Zr-based bulk metallic glass by nanosecond laser surface texturing. *J. Mater. Process. Technol.* **2020**, *283*, 116714. [CrossRef]
16. Huang, H.; Jun, N.; Jiang, M.; Ryoko, M.; Yan, J. Nanosecond pulsed laser irradiation induced hierarchical micro/nanostructures on Zr-based metallic glass substrate. *Mater. Des.* **2016**, *109*, 153–161. [CrossRef]
17. Li, L.; Hong, M.; Schmidt, M.; Zhong, M.; Malshe, A.; Huis In'Tveld, B.; Kovalenko, V. Laser nano-manufacturing—State of the art and challenges. *CIRP Ann. Manuf. Technol.* **2011**, *60*, 735–755. [CrossRef]
18. Chao, J.; Feng, J.; Chen, F.; Wang, B.; Tian, Y.; Zhang, D. Fabrication of superamphiphobic surfaces with controllable oil adhesion in air. *Colloids Surf. A Physicochem. Eng. Asp.* **2021**, *610*, 125708. [CrossRef]
19. Wang, Q.; Samanta, A.; Shaw, S.K.; Hu, H.; Ding, H. Nanosecond laser-based high-throughput surface nanostructuring (nHSN). *Appl. Surf. Sci.* **2020**, *507*, 145136. [CrossRef]
20. Lu, Y.; Guan, Y.C.; Li, Y.; Yang, L.J.; Wang, M.L.; Wang, Y. Nanosecond laser fabrication of superhydrophobic surface on 316L stainless steel and corrosion protection application. *Colloids Surf. A Physicochem. Eng. Asp.* **2020**, *604*, 125259. [CrossRef]
21. Wu, J.; Yin, K.; Xiao, S.; Wu, Z.; Zhu, Z.; Duan, J.A.; He, J. Laser Fabrication of Bioinspired Gradient Surfaces for Wettability Applications. *Adv. Mater. Interfaces* **2021**, *8*, 1–23. [CrossRef]

22. Yin, K.; Wu, Z.; Wu, J.; Zhu, Z.; Zhang, F.; Duan, J.-A. Solar-driven thermal-wind synergistic effect on laser-textured superhydrophilic copper foam architectures for ultrahigh efficient vapor generation. *Appl. Phys. Lett.* **2021**, *118*, 211905. [CrossRef]
23. Wang, Q.; Samanta, A.; Toor, F.; Shaw, S.; Ding, H. Colorizing Ti-6Al-4V Surface via High-Throughput Laser Surface Nanostructuring. *J. Manuf. Process.* **2019**, *43*, 70–75. [CrossRef]
24. Cheng, Y.; Song, J.; Dai, Y. Anti-reflective and anticorrosive properties of laser-etched titanium sheet in different media. *Appl. Phys. A* **2019**, *125*, 343. [CrossRef]
25. Yang, C.; Chao, J.; Zhang, J.; Zhang, Z.; Liu, X.; Tian, Y.; Zhang, D.; Chen, F. Functionalized CFRP surface with water-repellence, self-cleaning and anti-icing properties. *Colloids Surf. A Physicochem. Eng. Asp.* **2020**, *586*, 124278. [CrossRef]
26. Liu, Y.; Zhang, Z.; Hu, H.; Hu, H.; Samanta, A.; Wang, Q.; Ding, H. An experimental study to characterize a surface treated with a novel laser surface texturing technique: Water repellency and reduced ice adhesion. *Surf. Coat. Technol.* **2019**, *374*, 634–644. [CrossRef]
27. Wang, H.; Zhuang, J.; Qi, H.; Yu, J.; Guo, Z.; Ma, Y. Laser-chemical treated superhydrophobic surface as a barrier to marine atmospheric corrosion. *Surf. Coat. Technol.* **2020**, *401*, 126255. [CrossRef]
28. Ngo, C.V.; Chun, D.M. Control of laser-ablated aluminum surface wettability to superhydrophobic or superhydrophilic through simple heat treatment or water boiling post-processing. *Appl. Surf. Sci.* **2018**, *435*, 974–982. [CrossRef]
29. Chun, D.M.; Ngo, C.V.; Lee, K.M. Fast fabrication of superhydrophobic metallic surface using nanosecond laser texturing and low-temperature annealing. *CIRP Ann. Manuf. Technol.* **2016**, *65*, 519–522. [CrossRef]
30. Tran, N.G.; Chun, D.M. Simple and fast surface modification of nanosecond-pulse laser-textured stainless steel for robust superhydrophobic surfaces. *CIRP Ann.* **2020**, *69*, 525–528. [CrossRef]
31. Ngo, C.V.; Chun, D.M. Fast wettability transition from hydrophilic to superhydrophobic laser-textured stainless steel surfaces under low-temperature annealing. *Appl. Surf. Sci.* **2017**, *409*, 232–240. [CrossRef]
32. Gregorčič, P. Comment on "Bioinspired Reversible Switch between Underwater Superoleophobicity/Superaerophobicity and Oleophilicity/Aerophilicity and Improved Antireflective Property on the Nanosecond Laser-Ablated Superhydrophobic Titanium Surfaces". *ACS Appl. Mater. Interfaces* **2021**, *13*, 2117–2127. [CrossRef] [PubMed]
33. Dinh, T.-H.; Ngo, C.-V.; Chun, D.-M. Controlling the Wetting Properties of Superhydrophobic Titanium Surface Fabricated by UV Nanosecond-Pulsed Laser and Heat Treatment. *Nanomaterials* **2018**, *8*, 766. [CrossRef]
34. Lian, Z.; Xu, J.; Yu, Z.; Yu, P.; Ren, W.; Wang, Z.; Yu, H. Bioinspired Reversible Switch between Underwater Superoleophobicity/Superaerophobicity and Oleophilicity/Aerophilicity and Improved Antireflective Property on the Nanosecond Laser-Ablated Superhydrophobic Titanium Surfaces. *ACS Appl. Mater. Interfaces* **2020**, *12*, 6573–6580. [CrossRef] [PubMed]
35. Yong, J.; Chen, F.; Yang, Q.; Farooq, U.; Hou, X. Photoinduced switchable underwater superoleophobicity-superoleophilicity on laser modified titanium surfaces. *J. Mater. Chem. A* **2015**, *3*, 10703–10709. [CrossRef]
36. Wang, Q.; Wang, H.; Zhu, Z.; Xiang, N.; Wang, Z.; Sun, G. Switchable wettability control of titanium via facile nanosecond laser-based surface texturing. *Surf. Interfaces* **2021**, *24*, 101122. [CrossRef]
37. Jiao, Y.; Brousseau, E.; Han, Q.; Zhu, H.; Bigot, S. Investigations in nanosecond laser micromachining on the Zr52.8Cu17.6Ni14.8Al9.9Ti4.9 bulk metallic glass: Experimental and theoretical study. *J. Mater. Process. Technol.* **2019**, *273*, 116232. [CrossRef]
38. Jiao, Y.; Brousseau, E.; Nishio Ayre, W.; Gait-Carr, E.; Shen, X.; Wang, X.; Bigot, S.; Zhu, H.; He, W. In vitro cytocompatibility of a Zr-based metallic glass modified by laser surface texturing for potential implant applications. *Appl. Surf. Sci.* **2021**, *547*, 149194. [CrossRef]
39. Williams, E.; Brousseau, E.B. Nanosecond laser processing of Zr41.2Ti13.8Cu12.5Ni10Be22.5 with single pulses. *J. Mater. Process. Technol.* **2016**, *232*, 34–42. [CrossRef]
40. Du, C.; Wang, C.; Zhang, T.; Yi, X.; Liang, J.; Wang, H. Reduced bacterial adhesion on zirconium-based bulk metallic glasses by femtosecond laser nanostructuring. *Proc. Inst. Mech. Eng. Part H J. Eng. Med.* **2020**, *234*, 387–397. [CrossRef]
41. Huang, H.; Yan, J. Surface patterning of Zr-based metallic glass by laser irradiation induced selective thermoplastic extrusion in nitrogen gas. *J. Micromech. Microeng.* **2017**, *27*, 075007. [CrossRef]
42. Fornell, J.; Pellicer, E.; Garcia-Lecina, E.; Nieto, D.; Suriñach, S.; Baró, M.D.; Sort, J. Structural and mechanical modifications induced on Cu 47.5 Zr 47.5 Al 5 metallic glass by surface laser treatments. *Appl. Surf. Sci.* **2014**, *290*, 188–193. [CrossRef]
43. Zhang, D.; Cheng, Z.; Liu, Y. Smart Wetting Control on Shape Memory Polymer Surfaces. *Chem. A Eur. J.* **2019**, *25*, 3979–3992. [CrossRef] [PubMed]
44. Tanvir Ahmmed, K.M.; Grambow, C.; Kietzig, A.M. Fabrication of micro/nano structures on metals by femtosecond laser micromachining. *Micromachines* **2014**, *5*, 1219–1253. [CrossRef]
45. Feng, X.; Jiang, L. Design and creation of superwetting/antiwetting surfaces. *Adv. Mater.* **2006**, *18*, 3063–3078. [CrossRef]
46. Feng, L.; Li, S.; Li, Y.; Li, H.; Zhang, L.; Zhai, J.; Song, Y.; Liu, B.; Jiang, L.; Zhu, D. Super-hydrophobic surfaces: From natural to artificial. *Adv. Mater.* **2002**, *14*, 1857–1860. [CrossRef]
47. Gregorčič, P.; Šetina-Batič, B.; Hočevar, M.; Šetina-Batič, B.; Hočevar, M.; Šetina-Batič, B.; Hočevar, M. Controlling the stainless steel surface wettability by nanosecond direct laser texturing at high fluences. *Appl. Phys. A Mater. Sci. Process.* **2017**, *123*, 766. [CrossRef]
48. Fasasi, A.Y.; Mwenifumbo, S.; Rahbar, N.; Chen, J.; Li, M.; Beye, A.C.; Arnold, C.B.; Soboyejo, W.O. Nano-second UV laser processed micro-grooves on Ti6Al4V for biomedical applications. *Mater. Sci. Eng. C* **2009**, *29*, 5–13. [CrossRef]

49. Ngo, C.V.; Chun, D.M. Effect of Heat Treatment Temperature on the Wettability Transition from Hydrophilic to Superhydrophobic on Laser-Ablated Metallic Surfaces. *Adv. Eng. Mater.* **2018**, *20*, 1701086. [CrossRef]
50. Tian, Y.; Jiang, L. Wetting: Intrinsically robust hydrophobicity. *Nat. Mater.* **2013**, *12*, 291–292. [CrossRef]
51. Choi, H.J.; Min, B.H.; Shin, J.H.; Bae, D.H. Strengthening in nanostructured 2024 aluminum alloy and its composites containing carbon nanotubes. *Compos. Part A Appl. Sci. Manuf.* **2011**, *42*, 1438–1444. [CrossRef]
52. Akbarpour, M.R.; Salahi, E.; Hesari, F.A.; Yoon, E.Y.; Kim, H.S.; Simchi, A. Microstructural development and mechanical properties of nanostructured copper reinforced with SiC nanoparticles. *Mater. Sci. Eng. A* **2013**, *568*, 33–39. [CrossRef]
53. Wu, L.J.; Luo, K.Y.; Liu, Y.; Cui, C.Y.; Xue, W.; Lu, J.Z. Effects of laser shock peening on the micro-hardness, tensile properties, and fracture morphologies of CP-Ti alloy at different temperatures. *Appl. Surf. Sci.* **2018**, *431*, 122–134. [CrossRef]
54. Long, J.; Zhong, M.; Zhang, H.; Fan, P. Superhydrophilicity to superhydrophobicity transition of picosecond laser microstructured aluminum in ambient air. *J. Colloid Interface Sci.* **2015**, *441*, 1–9. [CrossRef]
55. Samanta, A.; Wang, Q.; Shaw, S.K.; Ding, H. Roles of Chemistry Modification for Laser Textured Metal Alloys to Achieve Extreme Surface Wetting Behaviors. *Mater. Des.* **2020**, *192*, 108744. [CrossRef]

micromachines

Article

PI Film Laser Micro-Cutting for Quantitative Manufacturing of Contact Spacer in Flexible Tactile Sensor

Congyi Wu [1,2], Tian Zhang [1,2], Yu Huang [1,2,*] and Youmin Rong [1,2,*]

1 State Key Lab of Digital Manufacturing Equipment and Technology, Huazhong University of Science and Technology, Wuhan 430070, China; cyw@hust.edu.cn (C.W.); tianz@hust.edu.cn (T.Z.)
2 School of Mechanical Science and Engineering, Huazhong University of Science and Technology, Wuhan 430070, China
* Correspondence: yuhuang_hust@163.com (Y.H.); rym@hust.edu.cn (Y.R.)

Abstract: The contact spacer is the core component of flexible tactile sensors, and the performance of this sensor can be adjusted by adjusting contact spacer micro-hole size. At present, the contact spacer was mainly prepared by non-quantifiable processing technology (electrospinning, etc.), which directly leads to unstable performance of tactile sensors. In this paper, ultrathin polyimide (PI) contact spacer was fabricated using nanosecond ultraviolet (UV) laser. The quality evaluation system of laser micro-cutting was established based on roundness, diameter and heat affected zone (HAZ) of the micro-hole. Taking a three factors, five levels orthogonal experiment, the optimum laser cutting process was obtained (pulse repetition frequency 190 kHz, cutting speed 40 mm/s, and RNC 3). With the optimal process parameters, the minimum diameter was 24.3 ± 2.3 μm, and the minimum HAZ was 1.8 ± 1.1 μm. By analyzing the interaction process between nanosecond UV laser and PI film, the heating-carbonization mechanism was determined, and the influence of process parameters on the quality of micro-hole was discussed in detail in combination with this mechanism. It provides a new approach for the quantitative industrial fabrication of contact spacers in tactile sensors.

Keywords: laser micro-cutting; PI film; contact spacer; tactile sensor

Citation: Wu, C.; Zhang, T.; Huang, Y.; Rong, Y. PI Film Laser Micro-Cutting for Quantitative Manufacturing of Contact Spacer in Flexible Tactile Sensor. *Micromachines* **2021**, *12*, 908. https://doi.org/10.3390/mi12080908

Academic Editors: Nam-Trung Nguyen and Lawrence Kulinsky

Received: 30 May 2021
Accepted: 28 July 2021
Published: 30 July 2021

1. Introduction

Flexible tactile sensors are widely used in electronic skin [1–3], robots [4–6], wearable devices [7–9], etc. Depending on the source of the signal, the tactile sensors are classified into capacitive-type [10,11], piezoelectric-type [12,13], triboelectric-type [14] and piezoresistive-type [15–18]. Among them, the piezoresistive tactile sensor has been extensively studied for its advantages of simple structure, low manufacturing cost and convenient signal processing. As one of the core components in piezoresistive tactile sensors, a contact spacer acts as a regulator of sensitivity and test range. For instance, a 2 μm thick rigid SiO_2 layer was applied as contact spacer and the sensor achieved ultra-high sensitivity of about 100–970 μA/kPa, but inflexibility limits its use scope [19]. In addition, a polyvinyl pyrrolidone (PVP) nanowire mesh was used to isolate the silver-plated micro-pyramids, creating a tactile sensor with an ultra-high sensitivity and ultra-wide detection range [20]. A tactile sensor with adjustable sensitivity and test range was fabricated by isolating the wrinkled polypyrrole (PVP) film through a polyvinyl alcohol (PVA) nanowire mesh [21]. However, the insulating nanowire mesh prepared by electrospinning process not only cannot quantify the size of single micro-hole, but also cannot uniformize the size of multiple micro-holes. The quantitative manufacturing of contact spacers thus should meet the following conditions: (1) flexible insulating film to isolate resistance, (2) ultra-thin film (1–10 μm) to form micro-sized peak in the micro-hole during compression deformation and (3) 5–100 μm diameter micro-hole can be fabricated quantitatively to form large contact resistance in a single micro-hole during compression deformation.

Polyimide (PI) film has many excellent properties such as flexibility, ultra-thin manufacturing (≥3 μm), insulation, high temperature resistance, radiation-resistance, etc., making it an excellent raw material for contact spacers. Ultra-thin PI film thus can be used to fabricate the contact spacer, and traditional machining methods such as stamping cannot fabricate micro-holes in the film. Laser-based micro-nano manufacturing has been widely used in various industries, such as metal organic framework (MOF) additive manufacturing [22], polydimethylsiloxane (PDMS) laser cutting in flexible electronics, laser direct writing for interdigitated electrode [23,24], laser induced biodegradation [25–27], etc. [28], laser micro-cutting thus may be a better method for preparing micro-holes through PI films. Nanosecond lasers are widely used in industrial production due to their low cost and high stability [29,30]. Ultraviolet (UV, 355 nm) laser has lower thermal ablation effect on the substrate than other wavelengths of light, so it is more suitable for manufacturing of organic materials to obtain smaller heat affected zones (HAZs) [31]. However, whether the interaction mechanism between the PI film and the UV nanosecond laser is a photothermal process or photochemical process, or both, remains to be further investigated [32–34]. In addition, the mechanism of PI film decomposition under laser irradiation conditions is still unclear, and the relationship between the thermal damage and the process parameters needs to be further investigated.

In this paper, nanosecond UV laser was chosen to fabricate the ultra-thin micro-hole PI film contact spacer. However, the evaluation index of micro-holes quality, the influence of process parameters on the evaluation index, and the interaction mechanism between UV laser and PI films are yet to be investigated. Firstly, an evaluation system of micro-hole based on roundness, diameter and size of the heat affected zone (HAZ) was established by analyzing the morphology after the laser micro-cutting. Secondly, the influence of process parameters was investigated by a 3-factor, 5-level orthogonal experiment to obtain the optimal process parameters. Finally, the interaction mechanism based on heating-carbonization was confirmed through the analysis of the products after laser micro-cutting and used to analyze the influence of process parameters. It provides a new approach for the quantitative industrial fabrication of contact spacers in tactile sensors.

2. Experiment Description

2.1. Laser Cutting System

As shown in Figure 1, the laser cutting system consisted of an industrial personal computer (IPC), nanosecond UV laser, extender lens, scan mirror, focusing lens and support platform. The nanosecond UV laser (Poplar-355-15A5, Huaray Precision Laser, Wuhan, China) had a maximum power of 12 W and pulse width < 15 ns. The extender lens was used to reduce laser processing power, because the diameter of laser spot expanded by the expander lens was larger than the entrance diameter of the scan mirror. The scan mirror (S10-355-D, SCANLAB, Pulheim, Germany) was used to control laser cutting path. The focusing lens (LINOS4401-402-000-20, QIOPTIQ, Gottingen, Germany) used in this study had a focal length of 167 mm. The 5 μm ultra-thin PI film (Taobao online store) was pasted on a piece of stainless steel, which was fixed on a motion platform. Output power of laser can be adjusted by setting pulse repetition frequency (PRF) values, and the correspondent relationship is shown in Table 1, where the power values are measured by power meter. After the laser processing, each sample was ultrasonically cleaned in deionized water and ethanol solution for 10 min in sequence, and then dried at room temperature.

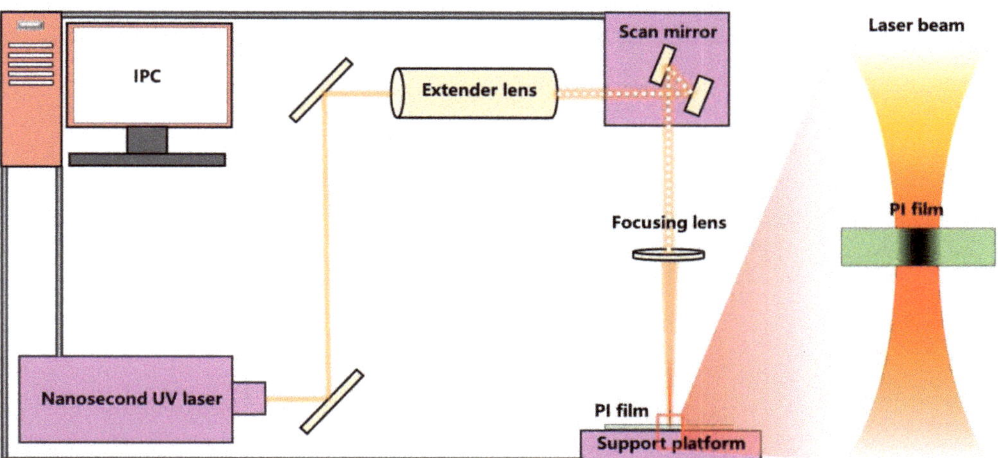

Figure 1. Experimental setup of PI film cutting process.

Table 1. Correspondent relationship between input PRF and output power.

Frequency (kHz)	100	120	140	160	180	200
Laser output power (W)	16.1	14.1	12.3	10.3	9.5	9.0
Actual cutting power (W)	0.159	0.137	0.121	0.102	0.94	0.89

2.2. Experiment Design of Laser Cutting Micro-Holes

According to previous experiments, the quality of micro-hole is mainly determined by the following three process parameters: PRF, cutting speed (CS) and repetition number of cuts (RNC). The nanosecond UV laser's PRF could be varied from 50 to 200 kHz, the CS could be varied from 0.1 to 10,000 mm/s, and the RNC could be increased indefinitely. Higher cutting powers, slower cutting speeds and more RNC help to remove material but can also lead to more defects such as large HAZ. Therefore, based on preliminary test results, the specific PRF value varied from 160 to 200 kHz, the CS value varied from 10 to 50 mm/s, and the RNC value varied from 1 to 5. Furthermore, each parameter was set to 5 arithmetic values within the range of variation, and their influences on the micro-cutting quality was evaluated through orthogonal experiment of 3 factors and 5 levels. The values and levels of the main process parameters are shown in Table 2, and the specific process parameters are shown in Table 3.

Table 2. Micro-cutting process parameters and levels.

Process Parameters	Unit	Notation	Factor Levels				
			1	2	3	4	5
Pulse repetition frequency	kHz	PRF	160	170	180	190	200
Cutting speed	mm/s	CS	10	20	30	40	50
Repetition number of cuts	——	CT	1	2	3	4	5

Table 3. $L_{25}(5^3)$ orthogonal experiments.

NO.	A	B	C	PRF	CS	RNC
1	1	1	1	160	0.01	1
2	1	2	2	160	0.02	2
3	1	3	3	160	0.03	3
4	1	4	4	160	0.04	4
5	1	5	5	160	0.05	5
6	2	1	2	170	0.01	2
7	2	2	3	170	0.02	3
8	2	3	4	170	0.03	4
9	2	4	5	170	0.04	5
10	2	5	1	170	0.05	1
11	3	1	3	180	0.01	3
12	3	2	4	180	0.02	4
13	3	3	5	180	0.03	5
14	3	4	1	180	0.04	1
15	3	5	2	180	0.05	2
16	4	1	4	190	0.01	4
17	4	2	5	190	0.02	5
18	4	3	1	190	0.03	1
19	4	4	2	190	0.04	2
20	4	5	3	190	0.05	3
21	5	1	5	200	0.01	5
22	5	2	1	200	0.02	1
23	5	3	2	200	0.03	2
24	5	4	3	200	0.04	3
25	5	5	4	200	0.05	4

2.3. Characterization

The basic morphology of the micro-hole was observed by a laser scanning confocal microscope (TCS SP8 X, Leica, Weztlar, Germany). In addition, a field emission scanning electron microscope (FS-EM, FEI Sirion 200, Santa Clara, CA, USA) was used to observe more detailed morphology and component analysis. The Raman spectra were carried out on a Raman spectrometer (HORIBA Jobin Yvon, Paris, France) with a 30 mW He-Cd laser of 532 nm. The thermogravimetric analysis (TGA) and differential thermal analysis (DTA) were measured by synchronous thermal analyzer (STA PT1600, LINSEIS, Zerb, Germany), that was carried out under air flow from room temperature to 800 °C at 10 min^{-1}.

3. Results and Discussion

3.1. Quality Evaluation System of Laser Micro-Cutting

Firstly, the evaluation system of micro-hole quality was established based on morphological characteristics, and the eigenvalues of each sample in the orthogonal experiment were extracted based on this system. Secondly, the influence of process parameters on micro-hole quality was discussed by orthogonal analysis of the eigenvalues of each sample. As shown in Figure 2, the typical micro-hole is not a perfect circle, and inscribed circle of micro-hole d_1 and circumscribed circle of micro-hole d_2 can be drawn. In addition, circumscribed circle d_3 can be drawn outside the micro-hole due to the presence of the HAZ. From the three-dimensional image of micro-hole measured by a laser scanning confocal microscope, it can be judged whether the micro-hole is through-hole (TH). According to the values of d_1, d_2, and d_3, evaluation indexes of the roundness, the diameter of the micro-hole, and the size of the HAZ can be determined, and the calculation formulas are as follows:

$$\text{Roundness} = \frac{d_2 - d_1}{2} \tag{1}$$

$$\text{Diameter} = \frac{d_1 + d_2}{2} \tag{2}$$

$$\text{Width of HAZ} = \frac{d_3 - \text{Dia}}{2} \quad (3)$$

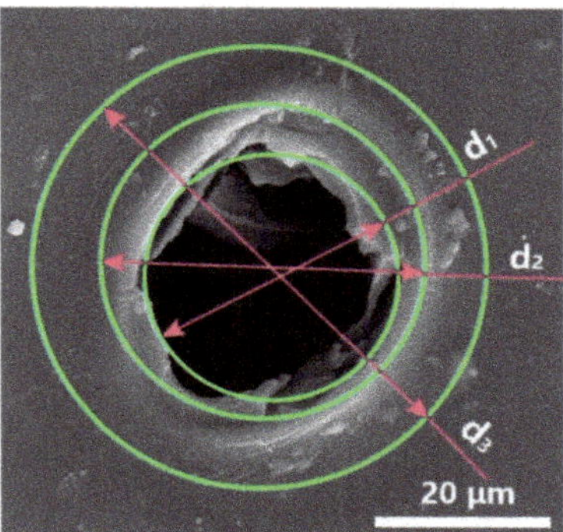

Figure 2. Schematic diagram of laser cutting quality evaluation.

Based on this evaluation system, eight samples were selected in each parameter for the calculation of the evaluation indexes. Sample values of d_1, d_2, d_3 were obtained by averaging five measurements after removing maximum and minimum values. In addition, it was determined that the cutting parameter can cut the PI film only when all the micro-holes were through-holes. The typical micro-hole morphologies of each parameter are shown in Figure 3, and the statistical data are shown in Table 4.

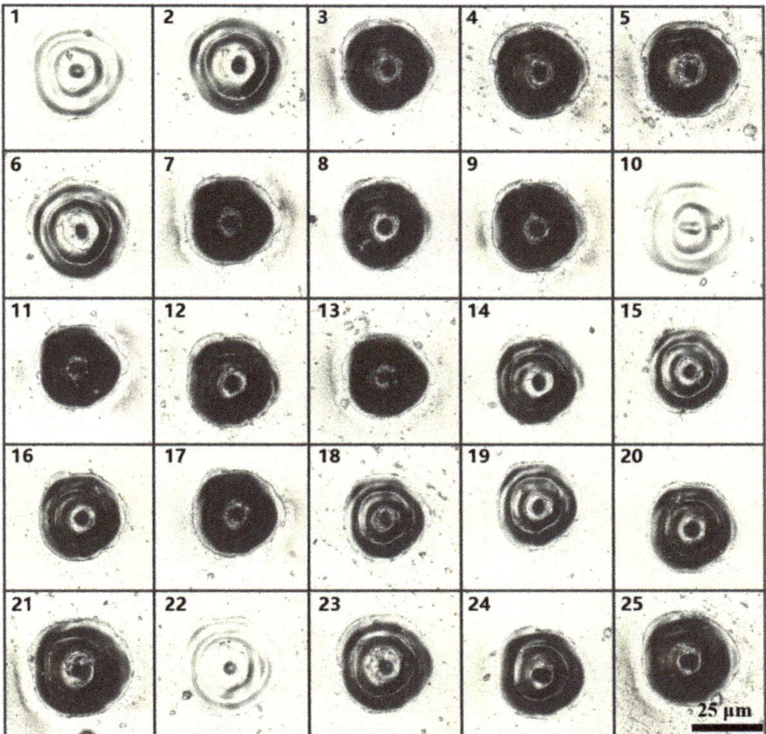

Figure 3. Typical confocal microscope image of each sample.

Table 4. Experimental results $L_{25}(5^3)$ orthogonal experiments.

NO.	PRF (kHz)	CS (mm/s)	RNC	Roundness (μm)	Diameter (μm)	Width of HAZ (μm)	TH
1	160	10	1	2.143	33.794	3.972	No
2	160	20	2	1.822	31.410	3.646	No
3	160	30	3	1.342	30.707	3.426	Yes
4	160	40	4	0.980	31.419	3.295	Yes
5	160	50	5	0.705	32.304	3.503	Yes
6	170	10	2	1.480	31.314	3.562	No
7	170	20	3	1.217	29.144	3.187	Yes
8	170	30	4	1.085	28.474	3.093	Yes
9	170	40	5	0.907	28.481	2.725	Yes
10	170	50	1	0.875	31.279	3.136	No
11	180	10	3	1.062	27.393	3.059	Yes
12	180	20	4	0.915	29.560	3.018	Yes
13	180	30	5	0.825	26.846	2.959	Yes
14	180	40	1	0.793	24.364	2.688	No
15	180	50	2	0.668	25.051	3.074	No
16	190	10	4	1.068	29.613	2.806	Yes
17	190	20	5	0.923	27.342	2.662	Yes
18	190	30	1	0.778	25.649	2.551	No
19	190	40	2	0.670	22.806	1.848	No
20	190	50	3	0.612	24.565	2.191	No
21	200	10	5	0.985	35.492	3.272	Yes
22	200	20	1	0.760	32.618	2.681	No
23	200	30	2	0.640	30.433	2.797	Yes
24	200	40	3	0.603	27.494	2.930	Yes
25	200	50	4	0.473	30.883	2.772	Yes

3.2. Variance Analysis of the $L_{25}(5^3)$ Orthogonal Experiments

Range is the difference between the maximum values and the minimum values of the experimental results. Range analysis can quickly determine the optimal level of a single factor and the optimal level combination of multiple factors in an orthogonal experiment. When the evaluation index is the roundness, the results of the range analysis of each parameter parameter are shown in Table 5 and Figure 4a. The parameter parameter PRF has a range R_1 of 0.706 μm for the evaluation index R. As the value of the PRF increases, the roundness value decreases continuously, and the best R is obtained at lever 5 (200 kHz). The range R_2 and R_3 of the process parameters CS and RNC are 0.681 and 0.201 μm, respectively. As the values of CS and RNC increase, the value of R also decreases, and the roundness is optimal at a cutting speed level of 5 (50 mm/s) and a cutting number level of 5 (5). Therefore, in this orthogonal experiment, the optimal R can be obtained when the parameter parameter combination is 5-5-5. As shown in Table 4, the through-hole can be obtained when the parameter combination is 5 (200 kHz) - 5 (50 mm/s) - 4 (4). Therefore, for the optimum parameter parameter combination 5(200 kHz) - 5(50 mm/s) - 5 (5), the through-hole can also be obtained in the case of only increasing RNC compared with the parameter parameter combination 5-5-4. In the range analysis of the orthogonal experiment, the larger the value of the range R, the greater the influence of the process parameters on the evaluation index. Comparing the values of R_1, R_2 and R_3, it is known that RNC, CS and PRF have an increased influence on the roundness of micro-hole in this orthogonal experiment.

Table 5. Range analysis table based on roundness evaluation.

	PRF (kHz)	CS (mm/s)	RNC
$\overline{K1}$	1.398	1.348	1.070
$\overline{K2}$	1.113	1.127	1.056
$\overline{K3}$	0.853	0.934	0.967
$\overline{K4}$	0.810	0.791	0.904
$\overline{K5}$	0.692	0.667	0.869
Optimal level	5	5	5
R_j	0.706	0.681	0.201
Order of range		$R_1 > R_2 > R_3$	

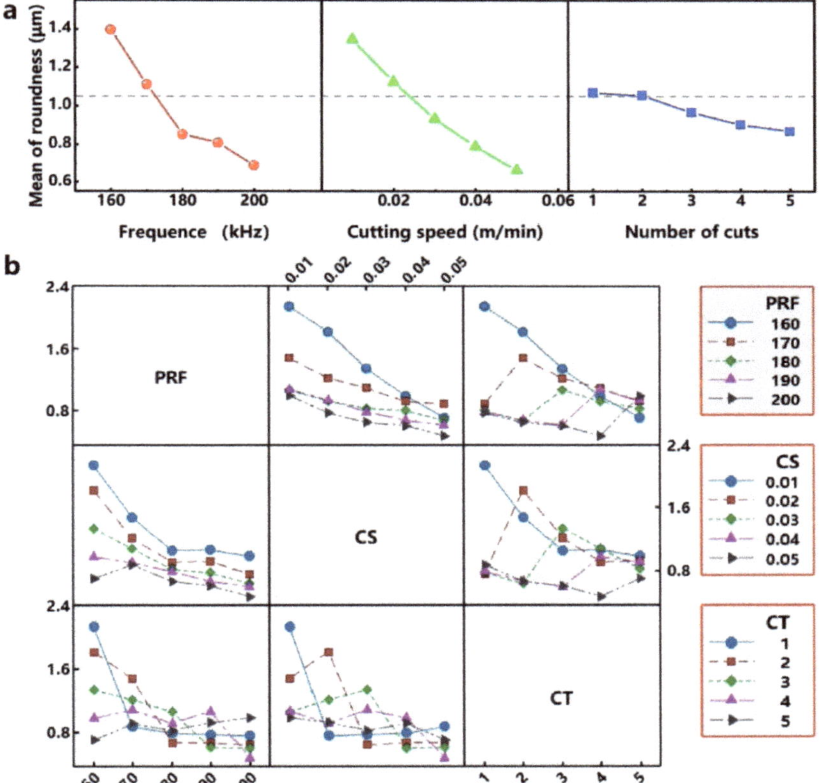

Figure 4. (**a**) Effect of process parameters on roundness. (**b**) Interactions between each parameter parameter for roundness.

The interactive diagram obtained through the interactive analysis can show how the relationship between a parameter parameter and an evaluation indicator depends on the value of the second parameter parameter. Figure 4b shows the interactions between each parameter parameter for roundness. The interaction analysis between the process parameters corresponding to the optimal roundness (200 kHz-50 mm/s-4) shows that (1) when the PRF level is 200 kHz, the interaction between CS and PRF is smaller than other PRF levels, and the interaction is smallest when the CS level is 50 mm/s. (2) When the PRF level is 200 kHz, the interaction between RNC and PRF is small compared to other PRF levels, but the interaction is greatest when the RNC level is 5. (3) When the CS level is 50 mm/s, the interaction between RNC and CS is smaller than the other CS level, and when the RNC level is 5, the interaction is larger than other RNC levels.

When the evaluation index is the diameter of the micro-hole, the results of the range analysis of each parameter parameter are shown in Table 6 and Figure 5a: (1) The parameter parameter PRF has a range R_1 of 5.932 μm, and the optimum parameter parameter level is 4 (190 kHz); (2) the parameter parameter CS has a range R_2 of 4.608 μm, and the optimum parameter parameter level is 4 (40 mm/s); (3) the parameter parameter RNC has a range R_3 of 1.680 μm, and the optimum parameter parameter level is 3 (3). As shown in Table 4, the through-hole can be obtained when the parameter parameter combination is 5 (200 kHz) - 4 (40 mm/s) - 3 (3). Therefore, for the optimum parameter parameter combination 4 (190 kHz) - 4 (40 mm/s) - 3 (3), the through-hole can also be obtained in the case of only reducing the output power compared with the parameter parameter combination 5-4-3. Comparing the values of R_1, R_2 and R_3, it is known that RNC, CS and PRF have an increased influence on the diameter of micro-hole in this orthogonal experiment. For the combination of process parameters (190 kHz - 40 mm/s - 3) corresponding to the optimal diameter of the micro-hole, it can be known from the interaction analysis in Figure 5b: (1) when the PRF level is 190 kHz, the interaction is smallest when the CS level is 40 mm/s. (2) When the PRF level is 190 kHz, the interaction is relatively small when RNC level is 3. (3) When the CS level is 40 mm/s, the interaction is middle when the RNC level is 3.

Table 6. Range analysis table based on diameter of the micro-hole.

	PRF (kHz)	CS (mm/s)	RNC
$\overline{K1}$	31.927	31.521	29.541
$\overline{K2}$	29.738	30.015	28.203
$\overline{K3}$	26.643	28.422	27.861
$\overline{K4}$	25.995	26.913	29.990
$\overline{K5}$	31.384	28.817	30.093
Optimal level	4	4	3
R_j	5.932	4.608	1.680
Order of range		$R_1 > R_2 > R_3$	

Similarly, when the evaluation index is the size of the HAZ, the results of the range analysis of each parameter parameter are shown in Table 7 and Figure 6a: (1) the parameter parameter PRF has a range R_1 of 1.157 μm, and the optimum parameter parameter level is 4 (190 kHz); the parameter parameter CS has a range R_2 of 0.637 μm, and the optimum process parameter level is 4 (40 mm/s); the parameter parameter RNC has a range R_3 of 0.047 μm, and the optimum parameter parameter level is 3 (3). As shown in Table 4, the through-hole can be obtained when the parameter parameter combination is 5 (200 kHz) - 4 (40 mm/s) - 3 (3). Therefore, for the optimum parameter parameter combination 4 (190 kHz) - 4 (40 mm/s) - 3 (3), the through-hole can also be obtained in the case of only reducing the output power compared with the parameter parameter combination 5-4-3. Comparing the values of R_1, R_2 and R_3, it is known that RNC, CS and PRF have an increased influence on the diameter of micro-hole in this orthogonal experiment. For the combination of process parameters (190 kHz - 40 mm/s - 3) corresponding to the optimal size of HAZ, it can be known from the interaction analysis in Figure 6b: (1) when the PRF level is 190 kHz, the interaction is smallest when the CS level is 40 mm/s. (2) When the PRF level is 190 kHz, the interaction is relatively small when RNC level is 3. (3) When the CS level is 40 mm/s, the interaction is relatively large when the RNC level is 3.

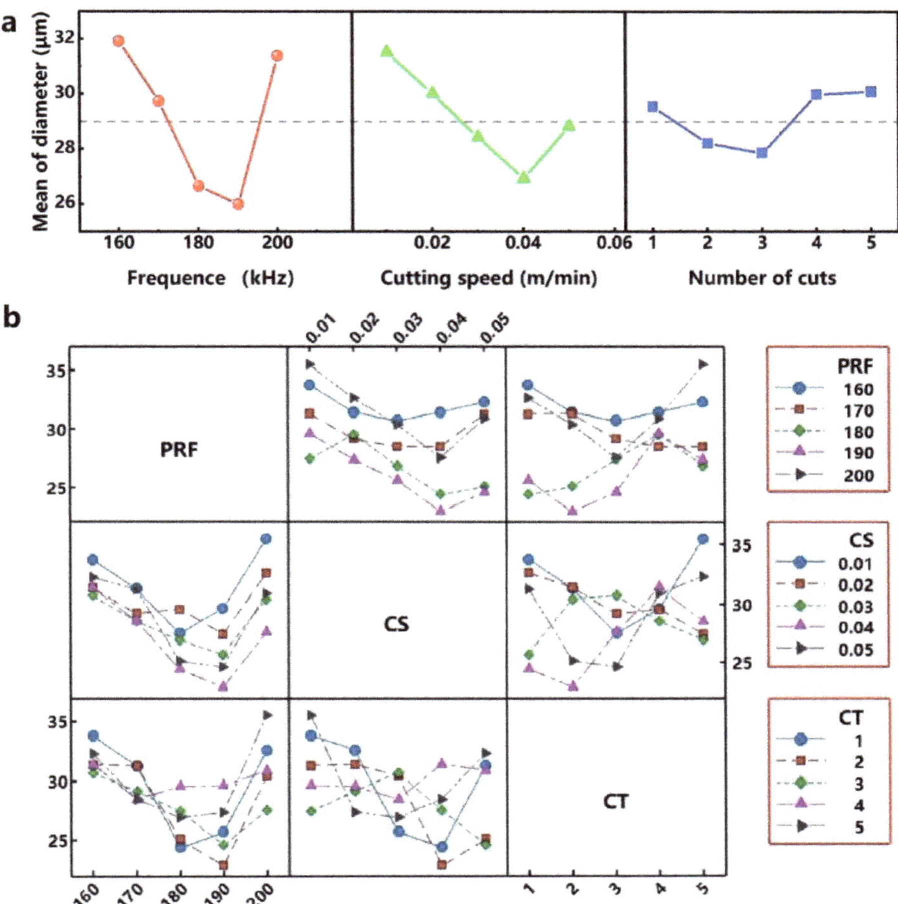

Figure 5. (**a**) Effect of process parameters on diameter of the micro-hole. (**b**) Interactions between each parameter parameter for diameter of the micro-hole.

Table 7. Range analysis table based on size of the HAZ.

	PRF (kHz)	CS (mm/s)	RNC
$\overline{K1}$	3.568	3.334	3.005
$\overline{K2}$	3.140	3.039	2.985
$\overline{K3}$	2.960	2.965	2.959
$\overline{K4}$	2.412	2.697	2.996
$\overline{K5}$	2.890	2.935	3.024
Optimal level	4	4	3
R_j	1.157	0.637	0.047
Order of range		$R_1 > R_2 > R_3$	

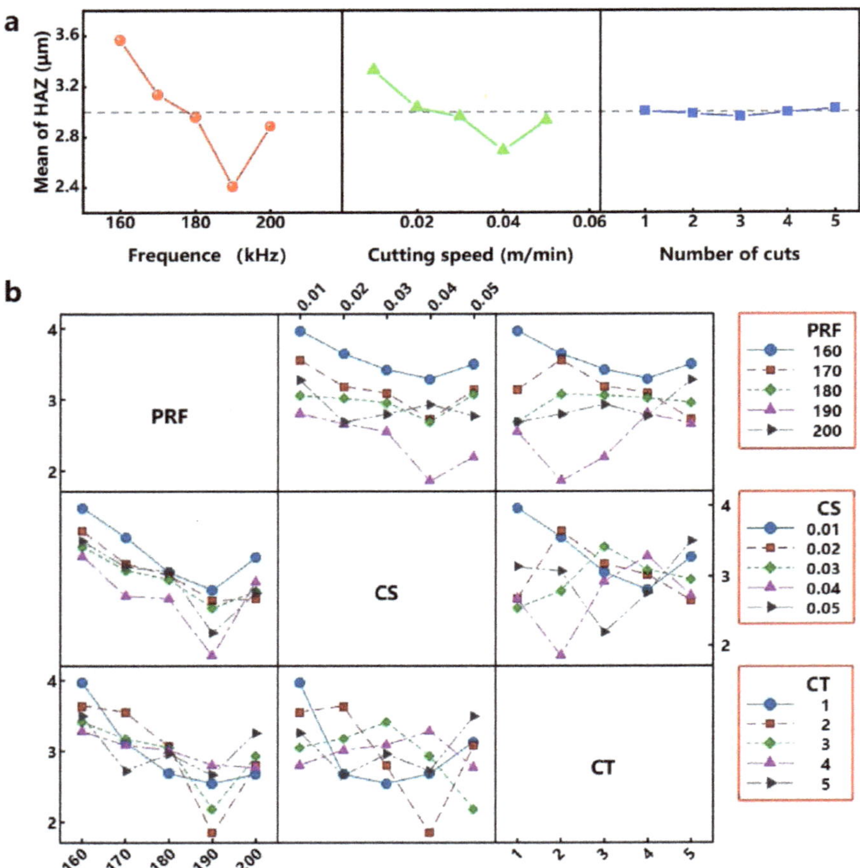

Figure 6. (**a**) Effect of process parameters on size of HAZ. (**b**) Interactions between each parameter parameter for size of HAZ.

According to the range analysis results of the orthogonal experiment, the effect of each parameter parameter on the diameter of micro-hole and HAZ was approximate, and the minimum diameter and HAZ were obtained at the same parameter level combination of 4(190 kHz) - 4(40 mm/s) - 3(3). As shown in Table 5, the roundness corresponding to each parameter parameter at this parameter level was 0.810 μm (PRF), 0.791 μm (CS), 0.976 μm (RNC), respectively. In the combination of parameter level to obtain the best roundness, the roundness corresponding to each parameter parameter was 0.692 μm (PRF), 0.667 μm (CS), 0.869 μm (RNC). Although the parameter level combination (5-5-5) that achieves the best roundness was inconsistent with this parameter level combination (4-4-3), the roundness difference of each parameter parameter was more than 0.13 μm. Therefore, parameter level combination of 4(190 kHz) - 4(40 mm/s) - 3(3) was selected as the optimal parameter level combination of laser micro-cutting in this study. It should be emphasized that in this parameter parameter combination, the interaction of PRF-CS and PRF-RNC is small, and the interaction of CS-RNC is large.

3.3. Mechanism of Laser Micro-Cutting

In order to determine how the process parameters affect the quality of the micro-holes, it is important to understand the interaction mechanism between the nanosecond UV laser and the ultra-thin PI film. Figure 7 shows the typical morphology of the micro-hole before and after cleaning. As shown in Figure 7a, after the laser micro-cutting, a circular area having a diameter of several tens of micrometers is formed through the PI film, the material in the micro-hole is not completely removed and a layered halo appears around this circular area. As shown in Figure 7b, the layered halo around the micro-hole disappeared after cleaning and the material in the micro-hole was removed, indicating that by-products of laser cutting could be removed.

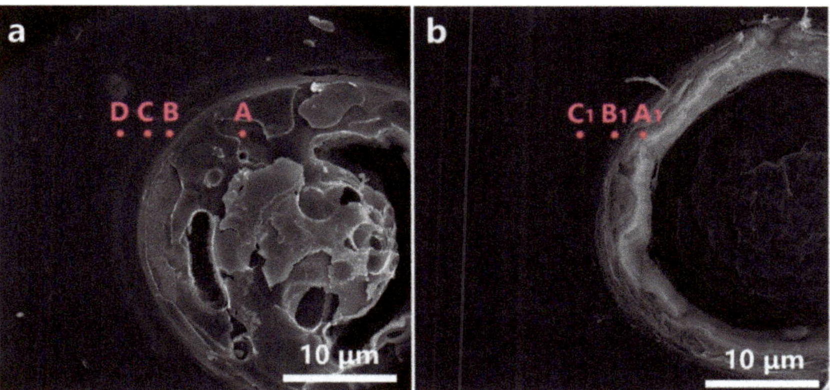

Figure 7. SEM images of the micro-hole (**a**) before cleaning and (**b**) after cleaning.

Further, the product of the laser micro-cutting was analyzed by Raman spectra and energy dispersion spectrum (EDS). Before cleaning, as shown in Figure 7a, the A-D points were selected from the inside of the micro-hole to the periphery of the layered halo for Raman analysis. There are no other peaks except the D band and the G band of point A at 1356 and 1594 cm^{-1} (Figure 8a), which indicates that the main component in the micro-hole after laser micro-cutting is carbon [35,36]. For the Raman spectra from point B to point D, there are no obvious peaks nor carbon peaks, which indicates that there are no new chemical bonds generated in the layered halo, and the carbon is not sputtered out from the micro-hole. According to the EDS analysis of these points, as shown in Table 8, the carbon content (97.99%) is much higher than the other positions, which further confirms the above-mentioned laser carbonization process.

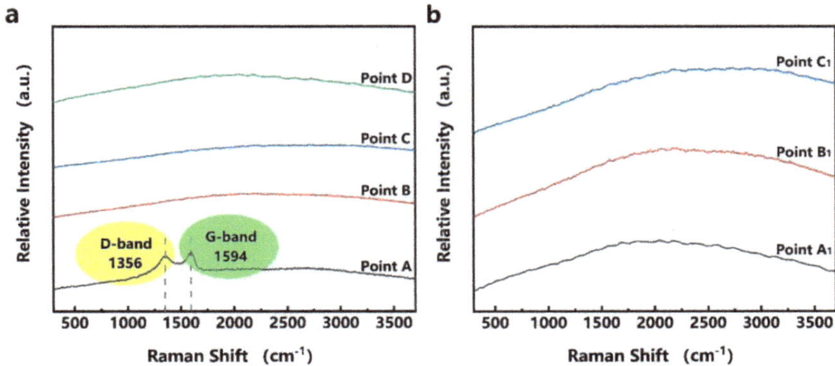

Figure 8. Raman measurements in different points in (**a**) Figure 7a and (**b**) Figure 7b.

Table 8. The atomic percentage in the designated area in Figure 7 before and after cleaning.

Atomic Percentage	C	O	N
Point A	97.99	0.86	1.15
Point B	73.37	19.51	7.12
Point C	81.88	12.93	5.19
Point D	71.64	20.94	7.42
Point A_1	73.44	19.38	7.18
Point B_1	70.87	20.69	8.44
Point C_1	72.58	19.56	7.86

After cleaning, as shown in Figure 7b, the point A_1 on the inner wall of the micro-hole, the point B_1 on the HAZ, and the point C_1 outside the HAZ were selected for Raman and EDS analysis. There are no obvious peaks or carbon peaks on the Raman curve in Figure 8b, indicating that the composition of each position is basically the same after cleaning. Points A_1, B_1, and C_1 have little difference in elemental content (Table 8), further indicating that there is no significant difference in the material surrounding the micro-hole after cleaning.

As shown in Figure 9, to analyze the dynamic change course of PI during laser heating, thermogravimetric analysis was performed on this ultra-thin PI film. According to the TGA curve, the plateau period is within 460 °C, and then the reaction period begins. Moreover, the thermal degradation process of PI film, a thermosetting material, is mainly one-step degradation. Thus, the weight loss step can be attributed to the carbonization process of the sample. As shown in the DTA curve, the temperature difference between the sample and the reference during heating (ΔT) is always less than zero, indicating that the PI film continues to absorb heat during thermal decomposition. The DTA curve has a peak at 619 °C, when the heat absorption rate is the highest and the weight loss rate is also the highest. In combination with the analysis of Figures 7 and 8 and Table 8, PI is carbonized at high temperature and remains on the inner wall, and the short-chain polyimide component is sputtered outside the micro-hole. In addition, because of thermal diffusion, wrinkles as shown in Figure 2 are produced around the micro-hole.

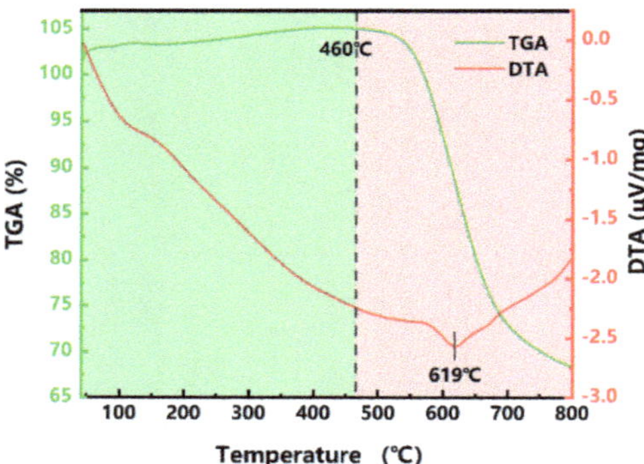

Figure 9. TGA and DTA curves of the ultra-thin PI film.

With heating-carbonization as the core, the influence of the laser micro-cutting process parameters on these evaluation indexes was discussed. The nanosecond UV laser carbonizes the PI film by high temperature on a circular cutting path to form a micro-hole, and the carbonized material remains in the micro-hole. Short-chain polyimide molecules that are partially decomposed but not yet charred will sputter on the outside of the micropores to form a halo. These products can be removed by cleaning, and a wrinkled HAZ formed by thermal diffusion can be clearly observed. Under the conditions of other process parameters unchanged: (1) the higher the PRF value, the lower the output power and the smaller the carbonization diameter of the spot; (2) the faster the CS, the less heat accumulation effect, and the smaller the carbonization diameter of the spot; (3) the more RNC, the more times the material is removed by the heat accumulation effect, and the larger the carbonization area. Experiment results of roundness analysis shows that lower output power and higher CS are beneficial to reduce the heat accumulation effect, which helps to improve the roundness of the micro-holes. In addition, more RNC is conducive to sufficient carbonized material in the same path, thus improving the roundness of the micro-hole too. Lower output power and faster cutting speeds help to form smaller carbonization diameter of the spot, resulting in smaller micro-hole diameters, but may also result in the inability to carbonize the material. The best process parameters should be achieved with a small carbonization diameter and fast cutting speeds, while orthogonal experiments help to quickly screen out the best combinations (4-4-3). At the same time, the smaller the diameter of the carbonized region, the smaller the area affected by thermal diffusion (HAZ), as shown in the experiment results of diameter and HAZ analysis of the micro-holes.

3.4. Fabrication of the Ultra-Thin PI Film Contact Spacer

As shown in Figure 10, an ultra-thin PI contact spacer with an average circle center distance of 50 μm was fabricated using the optimal parameter parameter combination (4-4-3). In the randomly selected 10 samples, the average and variance of the roundness, diameter, and HAZ are 0.6 ± 0.3, 24.3 ± 2.3, 1.8 ± 1.1 μm. The SEM morphology shows that the micro-holes are evenly distributed on the PI film, and there are no obvious defects, such as non-through holes and irregular shapes, etc.

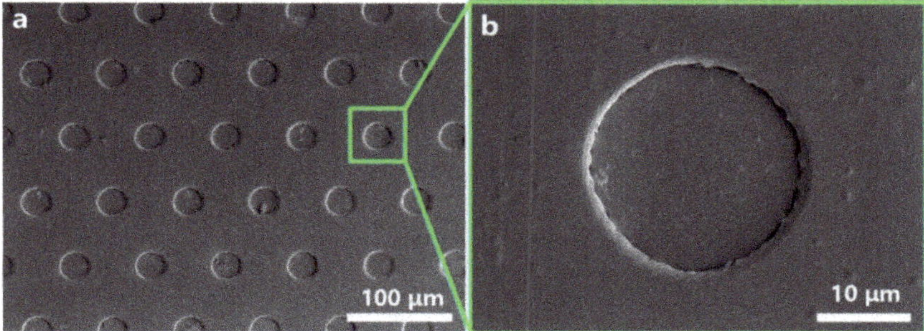

Figure 10. (**a**) Low- and (**b**) high-magnification SEM images of the ultra-thin PI film contact spacer.

4. Conclusions

The contact spacer is the core component that regulates the measurement range and sensitivity of tactile sensors. Using nanosecond UV laser commonly used in industry, the interaction mechanism between laser and ultra-thin PI film was first studied, and the influence of laser process parameters on micro-cutting quality was explored by an orthogonal experiment. Finally, the ultra-thin PI film contact spacer was successfully fabricated, which laid the foundation for the industrial production of the tactile sensor. Specific conclusions are as follows:

The high temperature generated by the spot carbonizes the PI film and partially stays in the micro-hole. The short-chain polyimide component is sputtered outside the micro-hole during the laser micro-cutting. Thermal diffusion during laser micro-cutting causes wrinkles around the micro-hole.

In the orthogonal experiment of this study, with the increase of PRF, CS and RNC values, the circularity of micro-hole was gradually optimized, and the optimum roundness was obtained at a parameter level of 5(200 kHz) - 5(50 mm/s) - 5 (5).

In the orthogonal experiment of this study, the effect of each parameter parameter on the diameter of micro-hole and HAZ was approximate. The minimum diameter (24.3 ± 2.3 μm) and HAZ (1.8 ± 1.1 μm) were obtained at the same parameter level of 4(190 kHz) - 4(40 mm/s) - 3(3).

Author Contributions: Data curation, T.Z.; funding acquisition, Y.H. and Y.R.; methodology, C.W.; writing—original draft, C.W.; writing—review and editing, C.W., Y.H. and Y.R. All authors have read and agreed to the published version of the manuscript.

Funding: This work was supported by the National Natural Science Foundation of China (51905191), the Major Project of Science and Technology Innovation Special for Hubei Province (2018AAA027), and Wuhan Science and Technology Planning Project (201903070311520).

Institutional Review Board Statement: Not applicable.

Informed Consent Statement: Not applicable.

Data Availability Statement: Not applicable.

Acknowledgments: The general characterization facilities are provided by the Flexible Electronics Manufacturing Laboratory in Experiment Center for Advanced Manufacturing and Technology in School of Mechanical Science and Engineering of HUST.

Conflicts of Interest: The authors declare no conflict of interest.

References

1. Nawrocki, R.A.; Matsuhisa, N.; Yokota, T.; Someya, T. 300-nm Imperceptible, Ultraflexible, and Biocompatible e-Skin Fit with Tactile Sensors and Organic Transistors. *Adv. Electron. Mater.* **2016**, *2*, 1500452. [CrossRef]
2. Chen, S.; Jiang, K.; Lou, Z.; Chen, D.; Shen, G. Recent Developments in Graphene-Based Tactile Sensors and E-Skins. *Adv. Mater. Technol.* **2018**, *3*, 1700248. [CrossRef]
3. Wu, C.; Tang, X.; Gan, L.; Li, W.; Zhang, J.; Wang, H.; Qin, Z.; Zhang, T.; Zhou, T.; Huang, J. High-Adhesion Stretchable Electrode via Cross-Linking Intensified Electroless Deposition on Biomimetic Elastomeric Micropore Film. *ACS Appl. Mater. Interfaces* **2019**, *11*, 20535–20544. [CrossRef]
4. Schneider, A.; Sturm, J.; Stachniss, C.; Reisert, M.; Burkhardt, H.; Burgard, W. Object Identification with Tactile Sensors using Bag-of-Features. In Proceedings of the 2009 IEEE/RSJ International Conference on Intelligent Robots and Systems, St. Louis, MO, USA, 10–15 October 2009.
5. Jamone, L.; Natale, L.; Metta, G.; Sandini, G. Highly Sensitive Soft Tactile Sensors for an Anthropomorphic Robotic Hand. *IEEE Sens. J.* **2015**, *15*, 4226–4233. [CrossRef]
6. Yuan, W.; Dong, S.; Adelson, E.H. GelSight: High-Resolution Robot Tactile Sensors for Estimating Geometry and Force. *Sensors* **2017**, *17*, 2762. [CrossRef]
7. Ahn, Y.; Song, S.; Yun, K.-S. Woven flexible textile structure for wearable power-generating tactile sensor array. *Smart Mater. Struct.* **2015**, *24*, 075002. [CrossRef]
8. Guo, X.; Huang, Y.; Cai, X.; Liu, C.; Liu, P. Capacitive wearable tactile sensor based on smart textile substrate with carbon black/silicone rubber composite dielectric. *Meas. Sci. Technol.* **2016**, *27*, 045105. [CrossRef]
9. Yeo, J.C.; Liu, Z.; Zhang, Z.Q.; Zhang, P.; Wang, Z.; Lim, C.T. Wearable mechanotransduced tactile sensor for haptic perception. *Adv. Mater. Technol.* **2017**, *2*, 1700006. [CrossRef]
10. Kwon, D.; Lee, T.-I.; Shim, J.; Ryu, S.; Kim, M.S.; Kim, S.; Kim, T.-S.; Park, I. Interfaces, highly sensitive, flexible, and wearable pressure sensor based on a giant piezocapacitive effect of three-dimensional microporous elastomeric dielectric layer. *ACS Appl. Mater. Interfaces* **2016**, *8*, 16922–16931. [CrossRef]
11. Qiu, Z.; Wan, Y.; Zhou, W.; Yang, J.; Yang, J.; Huang, J.; Zhang, J.; Liu, Q.; Huang, S.; Bai, N.; et al. Ionic skin with biomimetic dielectric layer templated from calathea zebrine leaf. *Adv. Funct. Mater.* **2018**, *28*, 1802343. [CrossRef]
12. Chen, Z.; Wang, Z.; Li, X.; Lin, Y.; Luo, N.; Long, M.; Zhao, N.; Xu, J.-B. Flexible piezoelectric-induced pressure sensors for static measurements based on nanowires/graphene heterostructures. *ACS Nano* **2017**, *11*, 4507–4513. [CrossRef]
13. Wu, W.; Wen, X.; Wang, Z.L.J.S. Taxel-addressable matrix of vertical-nanowire piezotronic transistors for active and adaptive tactile imaging. *Science* **2013**, *340*, 952–957. [CrossRef]
14. Fan, F.-R.; Lin, L.; Zhu, G.; Wu, W.; Zhang, R.; Wang, Z.L. Transparent triboelectric nanogenerators and self-powered pressure sensors based on micropatterned plastic films. *Nano Lett.* **2012**, *12*, 3109–3114. [CrossRef]
15. Zhu, B.; Niu, Z.; Wang, H.; Leow, W.R.; Wang, H.; Li, Y.; Zheng, L.; Wei, J.; Huo, F.; Chen, X.J.S. Microstructured graphene arrays for highly sensitive flexible tactile sensors. *Small* **2014**, *10*, 3625–3631. [CrossRef]
16. Pyo, S.; Lee, J.; Kim, W.; Jo, E.; Kim, J. Multi-Layered, Hierarchical Fabric-Based Tactile Sensors with High Sensitivity and Linearity in Ultrawide Pressure Range. *Adv. Funct. Mater.* **2019**, *29*, 1902484. [CrossRef]
17. Liu, W.; Liu, N.; Yue, Y.; Rao, J.; Cheng, F.; Su, J.; Liu, Z.; Gao, Y.J.S. Piezoresistive pressure sensor based on synergistical innerconnect polyvinyl alcohol nanowires/wrinkled graphene film. *Small* **2018**, *14*, 1704149. [CrossRef]
18. Tang, X.; Wu, C.; Gan, L.; Zhang, T.; Zhou, T.; Huang, J.; Wang, H.; Xie, C.; Zeng, D.J.S. Multilevel Microstructured Flexible Pressure Sensors with Ultrahigh Sensitivity and Ultrawide Pressure Range for Versatile Electronic Skins. *Small* **2019**, *15*, 1804559. [CrossRef] [PubMed]

19. Doostani, N.; Darbari, S.; Mohajerzadeh, S.; Moravvej-Farshi, M.K. Fabrication of highly sensitive field emission based pressure sensor, using CNTs grown on micro-machined substrate. *Sens. Actuators A Phys.* **2013**, *201*, 310–315. [CrossRef]
20. Luo, C.; Liu, N.; Zhang, H.; Liu, W.; Yue, Y.; Wang, S.; Rao, J.; Yang, C.; Su, J.; Jiang, X.J.N.E. A new approach for ultrahigh-performance piezoresistive sensor based on wrinkled PPy film with electrospun PVA nanowires as spacer. *Nano Energy* **2017**, *41*, 527–534. [CrossRef]
21. Wu, C.Y.; Zhang, T.; Zhang, J.; Huang, J.; Tang, X.; Zhou, T.; Rong, Y.; Huang, Y.; Shi, S.; Zeng, D. A new approach for an ultrasensitive tactile sensor covering an ultrawide pressure range based on the hierarchical pressure-peak effect. *Nanoscale Horiz.* **2019**, *4*, 1277–1285. [CrossRef] [PubMed]
22. Armon, N.; Greenberg, E.; Edri, E.; Kenigsberg, A.; Piperno, S.; Kapon, O.; Fleker, O.; Perelshtein, I.; Cohen-Taguri, G.; Hod, I.; et al. Simultaneous laser-induced synthesis and micro-patterning of a metal organic framework. *Chem. Commun. (Camb. Engl.)* **2019**, *55*, 12773–12776. [CrossRef]
23. Hsiao, W.-T.; Tseng, S.-F.; Huang, K.-C.; Chiang, D. Electrode patterning and annealing processes of aluminum-doped zinc oxide thin films using a UV laser system. *Opt. Lasers Eng.* **2013**, *51*, 15–22. [CrossRef]
24. Kim, H.; Auyeung, R.C.Y.; Lee, S.H.; Huston, A.L.; Pique, A. Laser-printed interdigitated Ag electrodes for organic thin film transistors. *J. Phys. D Appl. Phys.* **2010**, *43*, 085101. [CrossRef]
25. Antonczak, A.J.; Stepak, B.D.; Szustakiewicz, K.; Wojcik, M.R.; Abramski, K.M. Degradation of poly(l-lactide) under CO_2 laser treatment above the ablation threshold. *Polym. Degrad. Stab.* **2014**, *109*, 97–105. [CrossRef]
26. Stepak, B.; Antończak, A.J.; Szustakiewicz, K.; Pezowicz, C.; Abramski, K.M. The influence of ArF excimer laser micromachining on physicochemical properties of bioresorbable poly(L-lactide). In Proceedings of the SPIE LASE Laser-based Micro-& Nanoprocessing X, San Francisco, CA, USA, 13–18 February 2016.
27. Stepak, B.D.; Antonczak, A.J.; Szustakiewicz, K.; Koziol, P.E.; Abramski, K.M. Degradation of poly(L-lactide) under KrF excimer laser treatment. *Polym. Degrad. Stab.* **2014**, *110*, 156–164. [CrossRef]
28. Jayakumar, M.K.G.; Idris, N.M.; Zhang, Y. Remote activation of biomolecules in deep tissues using near-infrared-to-UV upconversion nanotransducers. *Proc. Natl. Acad. Sci. USA* **2012**, *109*, 8483–8488. [CrossRef]
29. Rong, Y.; Huang, Y.; Li, M.; Zhang, G.; Wu, C. High-quality cutting polarizing film (POL) by 355 nm nanosecond laser ablation. *Opt. Laser Technol.* **2021**, *135*, 106690. [CrossRef]
30. Wu, C.; Li, M.; Huang, Y.; Rong, Y. Cutting of polyethylene terephthalate (PET) film by 355 nm nanosecond laser. *Opt. Laser Technol.* **2021**, *133*, 106565. [CrossRef]
31. Wu, C.; Rong, Y.; Huang, Y.; Li, M.; Zhang, G.; Liu, W. Precision cutting of polyvinyl chloride film by ultraviolet nanosecond laser. *Mater. Manuf. Process.* **2021**. [CrossRef]
32. Lassithiotaki, M.; Athanassiou, A.; Anglos, D.; Georgiou, S.; Fotakis, C. Photochemical effects in the UV laser ablation of polymers: Implications for laser restoration of painted artworks. *Appl. Phys. A Mater. Sci. Process.* **1999**, *69*, 363–367. [CrossRef]
33. Schmidt, H.; Ihlemann, J.; Wolff-Rottke, B.; Luther, K.; Troe, J. Ultraviolet laser ablation of polymers: Spot size, pulse duration, and plume attenuation effects explained. *J. Appl. Phys.* **1998**, *83*, 5458–5468. [CrossRef]
34. Yingling, Y.G.; Garrison, B.J. Photochemical induced effects in material ejection in laser ablation. *Chem. Phys. Lett.* **2002**, *364*, 237–243. [CrossRef]
35. Tang, J.; Yamauchi, Y. Carbon materials: MOF morphologies in control. *Nat. Chem.* **2016**, *8*, 638–639. [CrossRef]
36. Zhang, L.; Ding, L.X.; Chen, G.F.; Yang, X.; Wang, H. Ammonia Synthesis under Ambient Conditions: Selective Electroreduction of Dinitrogen to Ammonia on Black Phosphorus Nanosheets. *Angew. Chem. Int. Ed.* **2019**, *131*, 2638–2642. [CrossRef]

Article

Medical Applications of Diode Lasers: Pulsed versus Continuous Wave (cw) Regime

Michał Michalik [1], Jacek Szymańczyk [2], Michał Stajnke [3], Tomasz Ochrymiuk [3] and Adam Cenian [3,*]

1 Centrum Medyczne MML, 00-112 Warszawa, Poland; michalim@mml.com.pl
2 Department of Dermatology, Medical University of Warsaw, 82a Koszykowa Street, 02-008 Warsaw, Poland; szymanczyk.jacek@gmail.com
3 Department of Physical Aspects of Ecoenergy, The Szewalski Institute of Fluid-Flow Machinery, Polish Academy of Sciences, 14 Fiszera Street, 80-952 Gdansk, Poland; michal.stajnke@imp.gda.pl (M.S.); tomasz.ochrymiuk@imp.gda.pl (T.O.)
* Correspondence: cenian@imp.gda.pl

Abstract: The paper deals with the medical application of diode-lasers. A short review of medical therapies is presented, taking into account the wavelength applied, continuous wave (cw) or pulsed regimes, and their therapeutic effects. Special attention was paid to the laryngological application of a pulsed diode laser with wavelength 810 nm, and dermatologic applications of a 975 nm laser working at cw and pulsed mode. The efficacy of the laser procedures and a comparison of the pulsed and cw regimes is presented and discussed.

Keywords: laser diodes; pulsed and continuous wave (cw) regimes; medical applications; dermatology; laryngology

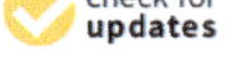

Citation: Michalik, M.; Szymańczyk, J.; Stajnke, M.; Ochrymiuk, T.; Cenian, A. Medical Applications of Diode Lasers: Pulsed versus Continuous Wave (cw) Regime. *Micromachines* **2021**, *12*, 710. https://doi.org/10.3390/mi12060710

Academic Editors: Congyi Wu and Yu Huang

Received: 5 May 2021
Accepted: 12 June 2021
Published: 17 June 2021

1. Introduction

We are approaching the 60th anniversary of laser medical applications. Shortly after the invention ruby lasers (with wavelength 694.3 nm) in the 1960s, Goldman et al. [1] started using it as therapy for melanoma, a human skin disease [2]. Later, in the 1980s, more powerful lasers, such as CO_2 lasers, argon lasers, and Nd:YAG lasers, were applied in the field of surgery (including laparoscopic), ophthalmology, dermatology, oncology, etc. An important step forward was the implementation of selective photothermolysis in dermatology by Anderson and Parrish [3], which are based on pigment-specific, short-pulsed lasers, e.g., Q-switched lasers.

Diode lasers (DLs), which first appeared in 1962, are still the most energy efficient and cost effective lasers. Therefore, they have found more and more applications in the field of medical therapies. Initially, DLs were not so popular as they gave power only in the order of mW. Diode lasers were used mainly for photobiomodulation (PBM)—previously also known as biostimulation or low-level laser therapy, LLLT— procedures, as well as for photodynamic therapy, where the wavelength is more crucial than high power [4,5]. Although PBM therapy was implemented by Endre Mester et al. [6] in 1967, for several decades it was mistrusted by many medical laser specialists. Only recently, after recognition of the role of cytochrome c oxidase in the mitochondrial respiratory chain as a primary chromophore and the introduction of the concept of "retrograde mitochondrial signaling", have attitudes changed. The significance of PBM in cell culture studies, resistance to fungal infections, mitigation of the side-effects of cancer therapy, pain and inflammation therapies, wound healing, muscle performance, etc. has become clearer. For example, Kowalec et al. [7] studied the Ceralas D15 diode laser which delivers optical power at 980 nm for wound and ulcer healing applications. The treatment enhanced wound healing and improved patient satisfaction and wellbeing. The study [8] confirmed the photomodulation efficacy of low power DL radiation at 740 nm (previously proven to be effective in wound healing) for therapy of dry eye disease. The radiation can improve corneal surface, alleviate inflammation through decreasing of the neutrophils levels, etc. Espey et al. [9] demonstrated that PBM using 665 nm pulsed (200 ns) radiation from a DL with fluence 4 and 6 J/cm^2 results in a significant increase in sperm motility and velocity within 120 min post-irradiation.

High power (above 5 W) DL applications for surgery have taken place in dermatology (see, e.g., [10]) and oral surgical procedures (e.g., [11]). Bass [10] demonstrated DLs ability to eliminate vascular lesions, in a therapy called photosclerosis or thermocoagulation, using a DL emitting 810 nm wavelength radiation and has achieved satisfactory effects without scarring. The applied laser fluence during the square wave pulse (5–15 ms) was 14–42 J/cm^2. The pulse interval was 32 ms (~31 Hz). Lesions treated included telangiectasias, spider veins, capillary dermal malformation and a cutaneous venous malformation. Telangiectasias were most responsive, usually disappearing after one treatment. Later, similar effects were achieved using a 980 nm diode laser by Desiate et al. [12]. Saetti et al. [13], after performing 22 endoscopic DL (810 nm) treatments of congenital subglottic hemangiomas, concluded that it is the safest and most effective (95% efficacy) therapy. The same cw DL radiation was used by Ferri et al. [14] to successfully treat Tis and T1 glottic carcinomas; all patients were able to eat without aspiration, as soon as the second day. Mittnacht et al. [15] applied DL radiation with power up to 450 W and λ = 808 and 940 nm to lung tissue. The laser with wavelengths 810 nm (3.5 W cw, 200 µm fiber) was used to treat oral Pyogenic granuloma [16].

More recently, Lee et al. [17] studied the efficacy of laser tonsillectomy using a 1940-nm laser working with a fluence of 12 W. The mean time for the procedure was 22.6 min and a notable reduction in pain at one week postoperative was elicited. Kang et al. [18] applied DL radiation with wavelength 1940 nm for the treatment of nasal congestion due to hypertrophied nasal turbinates. As the absorption coefficient of 1940-nm radiation in tissue is very high, the laser ablates tissue more precisely with less thermal damage. This clinical feasibility trial included eight patients with inferior turbinate hypertrophy. A rather low laser power of 4.5 W was applied leading to good medical results. In order to increase cutting efficiency of 940 nm DL, Agrawal et al. [19] studied the effect of various external chromophores (beetroot extract, erythrosine dye, hibiscus extract) applied on animal tissues. Staining of tissues with 3% erythrosine dye improved the efficacy of a 940 nm diode laser, by introducing sharper, wider cuts and clean incision with minimal charring when compared to beetroot, hibiscus, and saline chromophore.

In addition, the efficacy of 532-nm DL was investigated by treating a 50-year-old Korean female with oral erythro-leucoplakia [20]. Two months after the DL treatment, using a power of 6 W and 25 ms pulse, the operated region was well-healed without any significant scar contracture. Diode lasers emitting at wavelength 808 nm and different fluencies (12–14 J/cm^2) were tested for hair removal efficacy [21]; 30 ms laser pulses at a fixed rate of 7 Hz were applied. No significant difference was observed for both applied fluencies including patient comfort. The treatments were tolerated well without anaesthesia. The feasibility of a diode laser emitting at 1470 nm for blood vessel sealing was studied by Im et al. [22]. It was found that a power of 20 W and irradiation time of 5–10 s are adequate for effective sealing of blood vessels, although the higher power is required to cut the vessels.

Diode lasers with a central wavelength in the range 980 $\pm$ 10 nm have not been widely used up until now in high power clinical therapies. Romanos et al. [11] examined the wound healing after the application of a diode laser (980 nm) in oral surgical procedures, such as removal of soft tissue tumors, frenectomies, excision of gingival hyperplasias, vestibuloplasties, hemangioma removal, and periimplant soft tissue surgery. Laser radiation was applied both in pulsed and cw regime, with and without contact to the tissue. The advantages of this procedure were good coagulation properties; lack of bleeding, pain, scar tissue formation or swelling; and good wound healing. A few other examples related to otolaryngology procedures are known: turbinate reduction, nasal polypectomy, ablation of an oral papilloma, and photocoagulation of nasal telangiectasias [23]. Schmedt et al. [24]

has studied endovenous laser treatment of saphenous veins using a diode-laser emitting light of wavelength 980 nm which was transported via a 600 μm bare tipped optical fibre. Telangiectasias were most responsive, usually disappearing after one treatment [12]. Reynaud et al. [25] applied the 980 nm laser in laser-assisted lipolysis and Weiss et al. [26] in laser-assisted liposuction. Tunçel et al. [27] used DL (4–9 W) cw radiation to treat early glottic cancer and a year later Karasu et al. [28] applied DL radiation (3–5 W cw) to vocal fold polyps.

A Ceralas D15 diode laser delivering up to 15 watts of optical power at 980 nm using a quartz fiber delivery system was used to treat benign laryngeal lesions at office-based (outpatient) surgery—see [29]. Laser radiation (at power 12 W superpulse mode) was applied to a lesion through the working channel (3.7 mm in diameter) of the video fiberoptic esophagoscope. Some treated lesions such as: vocal polyps, leucoplakia, laryngeal hair showed significant improvement, yet required repeated procedures. On the other hand, patients with contact granuloma, subglottic stenosis and tracheal lesions showed partial remission with laser surgery. Recently, Karkos et al. [30,31] demonstrated the efficacy of a new "Π" surgical technique (using 980 nm DL laser, 4–9 W) postoperatively to improve quality of voice and swallowing. It was proven that the 980 nm diode laser appears to be safe and "friendly". Excellent long-term decannulation rates together with no significant deterioration in voice quality was achieved. Prażmo et al. [32] confirmed a positive effect of the repeated 980 nm laser pulsed irradiation (100 Hz) on intracanal *Enterococcus faecalis* biofilm elimination.

The effects of 975 nm radiation of dermatologic DL (in pulsed and cw mode) developed in IMP PAN was studied using optical phantoms of skin [33] before its clinical application [34]. Further research comparing the interaction effects of radiation at 532, 975, and 1064 nm was performed and reported by Milanic et al. [35]. It was concluded that the risks of the epidermis or subcutaneous tissue overheating are significantly reduced.

The aim of this paper is to describe and analyse the medical application of diode lasers operating in pulsed and cw regimes, with a special focus on laryngological or dermatological therapies. The results related to the authors' experience in the field are presented and discussed, including first simulations of dermatologic treatment. The advantage of pulsed laser application is discussed and its limitations are analysed.

2. Materials & Methods

The medical therapies analysed here were performed for several hundred patients treated in a private clinic, the Medical Centre MML in Warsaw (in the field of laryngology) and the private dermatology practice of Dr J. Szymańczyk, in cooperation with the Institute of Fluid-Flow Machinery PAS in Gdańsk.

The Institute developed a dermatologic diode laser emitting at 975 nm, working at cw or pulsed regime—pulse lengths 100 ns − 300 ms, and laser output power 20 W [36], which was later implemented for therapies of neurofibroma and hemangiomas [34]. The second diode laser applied in MML Centre generated radiation with wavelength 810 nm and a pulse duration 4 s. In both cases, high efficacy of laser treatments was registered. Efficacy of procedure was defined as the ratio of the number of patients with positive effects of treatment therapy to the total number of procedures performed.

Besides medical treatments, the theoretical modelling and analyses of laser radiation interaction with neurofibroma blisters were performed. Therefore, the classic fluid-solid interaction problem is simulated and solved, in which the use of the monolithic method [37] is justified. First of all, non-trivial coupling of the thermal-FSI type [38] is considered, with the laser beam providing a heat stream to the tumor surface. There is an unstable flow of heat stream through various types of tissues to the tumor interior filled with fluid. This fluid heats up and there is a phase change, combined with a rapid increase in pressure, which results in a significant non-linear increase in tumor volume due to the hyperelastic properties of the skin. The tumor eventually explodes after some time, less than the time it takes to reach the pain threshold. The Arbitrary Lagrangian-Eulerian (ALE) description gives a proper foundation for monolithic methods in which simultaneous solution for all unknowns of the coupled fluid/solid system [39] and all interaction effects between the dependent equations are included. The set of balance equations in the well-known ALE form [40,41] are solved

$$\frac{\partial}{\partial t}\begin{Bmatrix} \rho \\ \rho \mathbf{v} \\ \rho e \end{Bmatrix} + \mathrm{div}\begin{Bmatrix} \rho \mathbf{v} \\ \rho \mathbf{v} \otimes \mathbf{v} \\ \rho e \mathbf{v} \end{Bmatrix} = \mathrm{div}\begin{Bmatrix} 0 \\ \mathbf{t} \\ \mathbf{t}\mathbf{v} + \mathbf{q} \end{Bmatrix} + \begin{Bmatrix} 0 \\ \rho \mathbf{b} \\ \rho \mathbf{b}\mathbf{v} \end{Bmatrix}, \tag{1}$$

where ρ is the density of the continuum particle, $\mathbf{v}$ is velocity of the continuum particle, $e = c_v T + \frac{1}{2}\mathbf{v}^2$ is total energy, c_v is specific heat at constant volume, T is temperature of the continuum particle, $\mathbf{t}$ is the Cauchy stress flux, $\mathbf{q} = \lambda \cdot T\nabla$ is the molecular heat flux defined by Fourier law (λ is thermal conductivity coefficient), and $\mathbf{b}$ is the earth acceleration. The Cauchy stress flux can be divided into an elastic part and a diffusive part:

$$\mathbf{t} = \mathbf{P} + \boldsymbol{\tau}^c, \tag{2}$$

where $\mathbf{P}$ is an elastic momentum flux which is reversible and $\boldsymbol{\tau}^c$ is a total diffusive momentum flux which describes irreversible phenomena. Below the first introductory results of simulations are presented and analyzed.

3. Results

Here, the results of diode laser treatments performed in MML Centre (laryngology) and a private dermatology practise are presented and discussed.

3.1. Pulsed Diode Laser 810 nm (5 W Fluence and Pulse Duration 4 s) in Laryngology Applications in MML Centre

(i) Laser-assisted somnoplasty using the palisade technique, a method of treatment for snoring and sleep apnea, is characterized by high efficacy, a short recovery period, and minimal risk of complications [42]. The method is implemented for palatoplasty, surgery of palatoglossal and palatopharyngeal arch, and uvuloplasty. During the procedure, the diode laser fibre is introduced into the soft palate (see Figure 1), which results in the formation of linear intra-parenchymal adhesions that stiffen the palate and shift it in the vertical plane. The therapy results in the prevention of tissue vibration during sleep, which, in turn, leads to increased sleep comfort and maximally widened airways. There are several advantages for application of this laser-assisted procedure, e.g., it enables a shorter surgery time (30–40 min), under local anaesthetic conditions. Shortly after the procedure, the patient can be discharged.

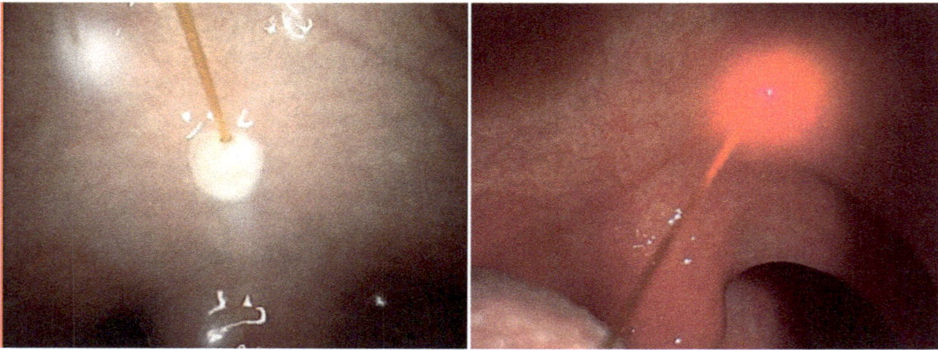

Figure 1. Introduction of laser fiber into a soft palate.

In years 2007–2020, 84 diode laser-assisted somnoplasty procedures using palisade technique were performed. Complete clinical response was observed in 77 cases, and a partial response was seen in seven cases. The efficacy of the therapy reached 92%.

(ii) Separation of adhesions in nasal septum is needed due to postoperative complications—see Figure 2. The adhesions being postoperative (iatrogenic) cicatrix appear between nasal conches and septum and inhibit normal air flow. After laser assisted separation, instead of the usual tamponade, a gel dressing, which dissolves after a certain period, is applied as sufficient. The laser procedure is safer for the patient and gives better results. From 2007 to 2020, 51 laser-assisted separation procedures were performed. Complete clinical response was observed in 49 cases, and a partial response was seen in two cases. The efficacy of the therapy reached 97%.

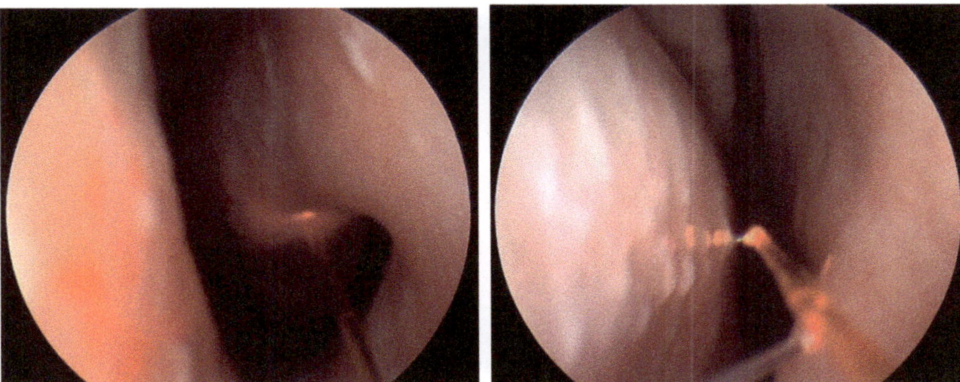

Figure 2. Laser assisted separation of adhesions in nasal septum.

(iii) Laser assisted frenuloplasty, a surgery for a short frenulum and frenectomy of labial frenulum is a simple, sensitive and safe medical procedure (Figure 3a). It is preceded by a local anaesthesia. The diode laser assisted therapy is bloodless and painless due to the character of laser radiation tissue interaction (increased coagulation). During the period 2007–2020, 62 diode laser-assisted frenectomy procedures were performed. Complete clinical response was observed in 61 cases, and a partial response was seen in one case, giving a procedure efficacy of 98%.

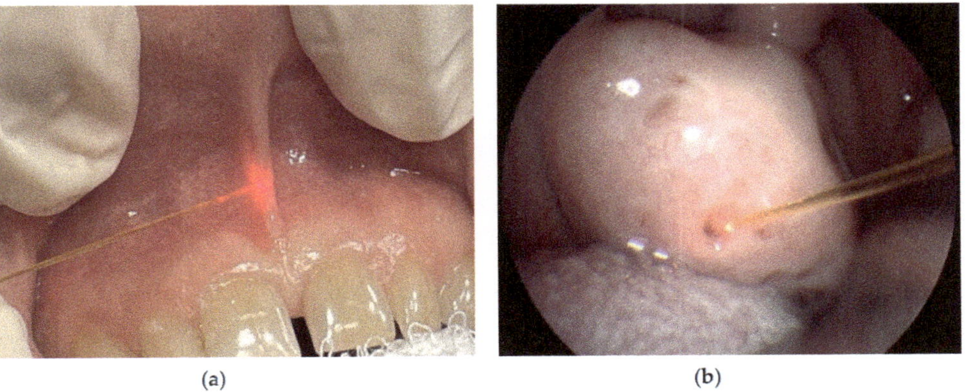

(a) (b)

Figure 3. Frenectomy of labial frenulum (**a**) and laser-assisted closure of tonsillar crypts (**b**).

(iv) Laser-assisted closure of tonsillar crypts after removal of debris (known as tonsil stones) resulting from bacterial and viral infections (see Figure 3b). After the debris removal a diode laser fiber is introduced, which enables shrinking and closing of crypts. This is an ambulatory (also known as office-based or Outpatient) procedure under local anaesthetic, and is painless and bloodless. During the period 2007–2020, 31 diode laser-assisted closures of tonsillar crypts were performed. Complete clinical response was observed in 29 cases, and a partial response was seen in two cases, giving an efficacy of 96%.

(v) Laser-assisted haemostasis (coagulation) results from interaction of 810 nm radiation of diode laser with the blood and lymphatic vessels—see Figure 4a. The process enables bloodless procedures and eliminates haemorrhaging both during and postoperatively. The process efficacy reaches 100%.

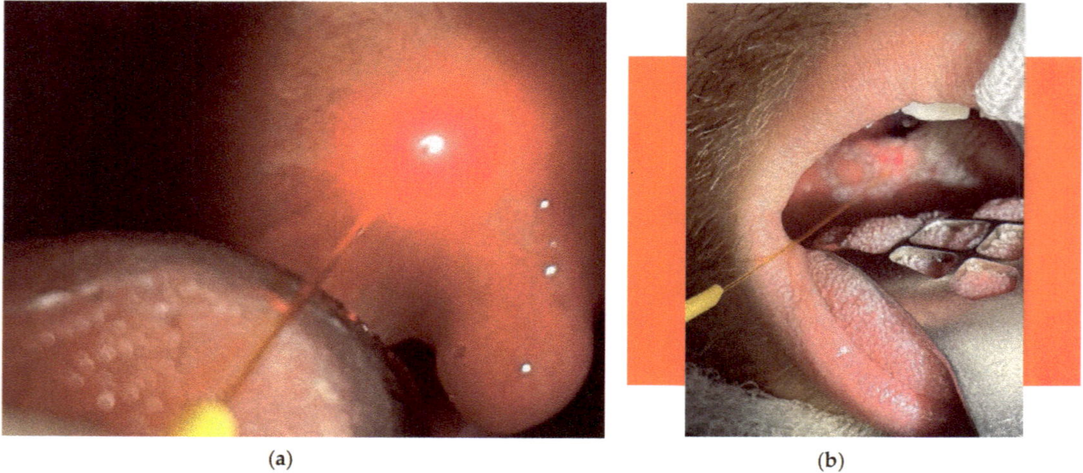

(a) (b)

Figure 4. Laser-assisted (**a**) haemostasis and (**b**) surgery of laryngopharynx.

(vi) Laser surgery of laryngopharynx and larynx (Figure 4b) enables sensitive and precise operation, removal of deteriorated tissues and protection of healthy ones. The separated tissue can be sent for histopathologic diagnostics. During the period 2007–2020, 54 laser surgery procedures were performed. Complete clinical response was observed in 50 cases, and a partial response was seen in 4 cases, resulting in an efficacy for the procedure of 93%.

(vii) Laser-assisted removal of cancerous changes/tissues (papilloma, polyps, haemangiomas, vocal nodules) enables precise operation and reaching narrow channels in nasal, sinus and other regions—see Figure 5. There is a low risk of thermal damage to tissue, so introduced wounds normally heal fast. The procedures are relatively fast and less invasive than standard ones. The laser haemostasis inhibits haemorrhage. During the period 2007–2020, 67 diode laser-assisted removals of cancerous changes were performed. Complete clinical response was observed in 64 cases, and a partial response was seen in three cases, giving an efficacy for the procedure of 95%.

(viii) The laser-assisted blepharoplasty (popular cosmetic eyelid surgery) is a medical/cosmetic procedure leading to correction of upper eyelid drooping (Figure 6a). It consists in removal of skin surplus from the upper eyelid. The procedure enables an increase of eyeshot (improved field of vision) and face rejuvenation. Its efficacy reaches 99%. During period 2007–2020, 97 diode laser-assisted blepharoplasty procedures were performed. Complete clinical response was observed in 96 cases, and a partial response was seen in one case.

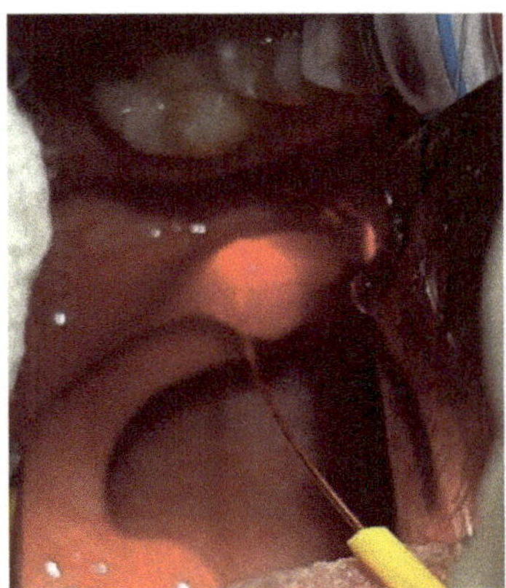
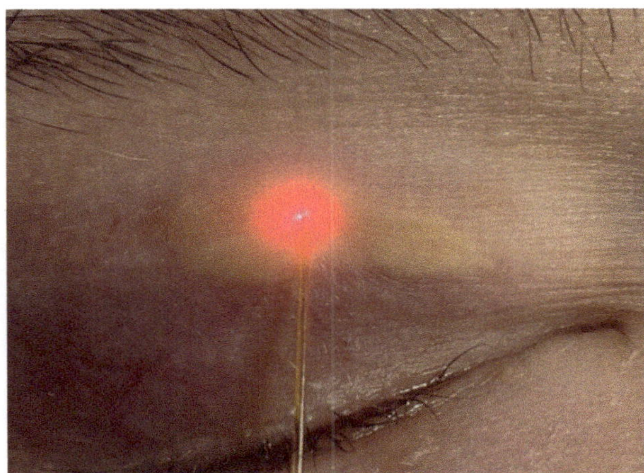

Figure 5. Laser-assisted removal of cancerous tissue.

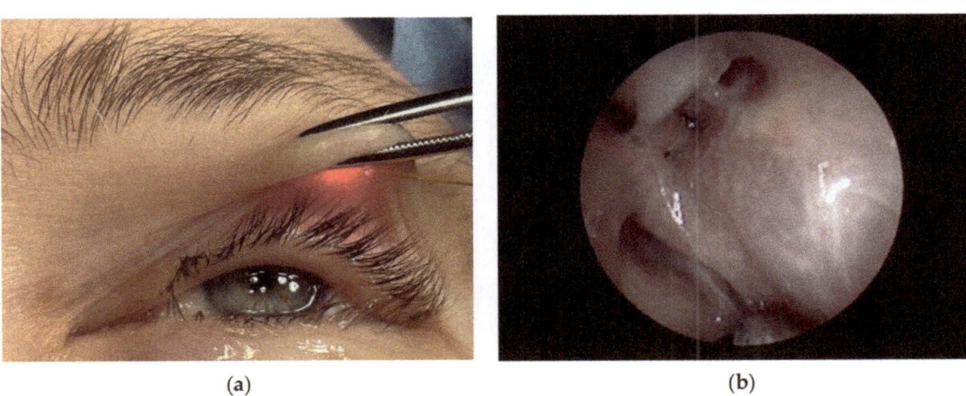

Figure 6. The laser-assisted (**a**) blepharoplasty and (**b**) nasolacrimal duct anastomosis 12 months after procedure.

(ix) Laser assisted dacryocystorhinostomy (DCR) was performed using a diode laser 810 nm, at power 8–10 W and pulses 0.5–1 s, in the case of patients with tear duct obstruction [43]. An elastic laser fiber 0.4 mm wide was introduced through the tear duct towards the lacrimal sac. The procedure was performed for 60 patients (44 women and 16 men) with average age 60.9 years. Positive effects were observed in the case of 96%, 75%, and 78%, after three, six, and 12 months, respectively (Figure 6b). In two cases the procedure was repeated and general efficacy increased to 81%. The intraoperative use of mitomycin C during the procedure of nasolacrimal duct anastomosis with diode laser increases its effectiveness [44].

Summing up, the utilisation of a 810 nm diode laser allows not only the removal of damaged tissue but it leads to haemostasis of blood vessels which in turn results in higher safety of therapies. This is of special importance when dealing with blood engorged tissues, where the risk of postoperative complications can be much higher.

3.2. Pulsed and Continuous (cw) Operation Regime of Diode Laser 975 nm Implemented for Therapy of Dermal Neurofibroma

In the case of patients affected by *dermal neurofibroma* disease, therapy proceeded at different levels of laser power in order to find the optimum conditions. Figure 7 presents the effects in the case, when laser radiation with cw power 10 W and 15 W pulsed regime (pulse 50 ms, 10 Hz) was applied to treat right side of the décolleté area. In the second case (see Figure 8), cw power 12 W was applied. The check after ~4–9 weeks have shown that the best therapeutic and cosmetic results have been achieved for cw power of 10 W. In the case of higher powers the healing period was longer as well as the cosmetic effect less desirable due to tendency to scarring.

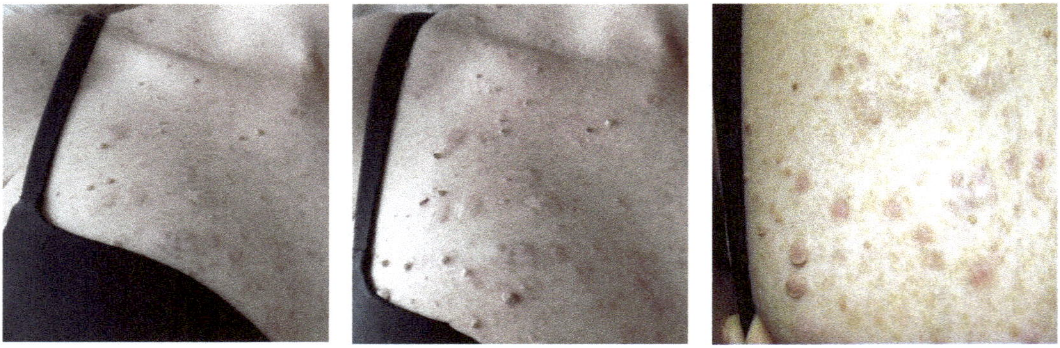

(a) (b) (c)

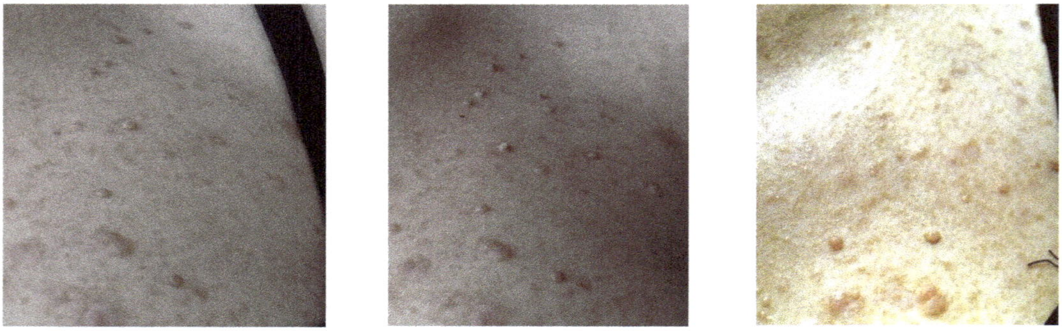

(d) (e) (f)

Figure 7. Laser therapy of dermal neurofibroma at right side of the décolleté area using DL radiation with wavelengths 975 nm: and cw power 10 W; (**a**) view before irradiation, (**b**) soon after irradiation (**c**) 7 weeks after laser treatment; and with pulsed power 15W (pulse 50 ms, 10 Hz) left side of the décolleté area (**d**) view before irradiation, (**e**) soon after irradiation (**f**) 7 weeks after laser treatment.

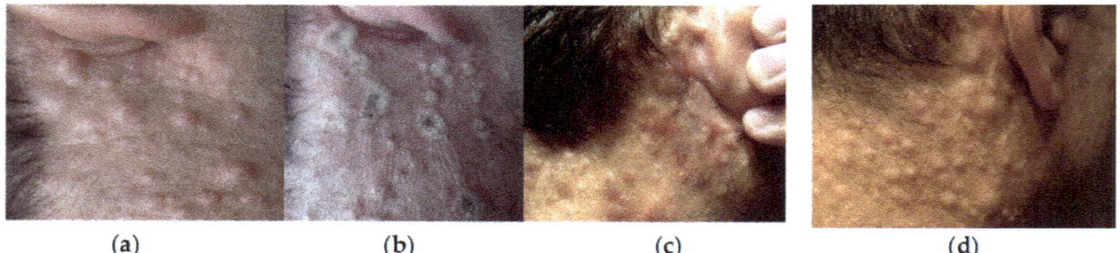

(a) (b) (c) (d)

Figure 8. Laser therapy of dermal neurofibroma at right side of the neck below ear region using DL radiation with wavelengths 975 nm and continues wave, power: 12 W; (**a**) view before irradiation, (**b**) soon after irradiation (55 s), (**c**) 7 weeks after laser treatment, (**d**) one year after therapy.

Application of lower radiation powers does not significantly improve the final therapeutic or cosmetic effect, i.e., by flattening of irradiated distortions or reduced tendency to scarring. It looks like the applied pulsed regime gives slightly better results (compare Figure 7c,f), the procedure is slightly less painful and better tolerated by patients. However, the procedure lasted longer. In the case of the patient presented in Figure 8, the effects one year after therapy may point to the need for therapy repetition.

Due to dolorability of the therapy using both diode (975 nm), Nd:YAG and Ho:YAG lasers the treatment was preceded by local anaesthesia with 1% of lignocaine. However, patients' reactions and tolerance of these laser therapies were variable. In the case of diode laser (975 nm) patients do not experience real pain or any tissue warming despite deep coagulation. The tissue coagulation proceeds fast and effectively. The reaction to Ho:YAG laser irradiation (2100 nm) was different. Patients despite local anaesthesia very often suffered unpleasant high temperature effects due to the laser irradiation and coagulation process. The treatment (Ho:YAG laser irradiation) of the skin, necessary to achieve the required result, lasts significantly longer than in the case of the laser diode.

In order to understand better the phenomena and mechanism of neurofibroma therapy, a theoretical modelling and analyses of laser radiation interaction with neurofibroma blister was performed. As mentioned in Section 2 the fluid contained in these cancerous blisters is heated by laser radiation and evaporates rapidly increasing pressure. The blister eventually explodes after some time, e.g., at least 3 s for blisters of 1.8 mm in diameter and more than 5 s for blisters 3 mm in diameter. These results correlate well with the results of introductory simulations based on the ALE model described above, e.g., the full evaporation of liquid in the neurofibroma blister occurred after 3 s of laser heating—see Figure 9.

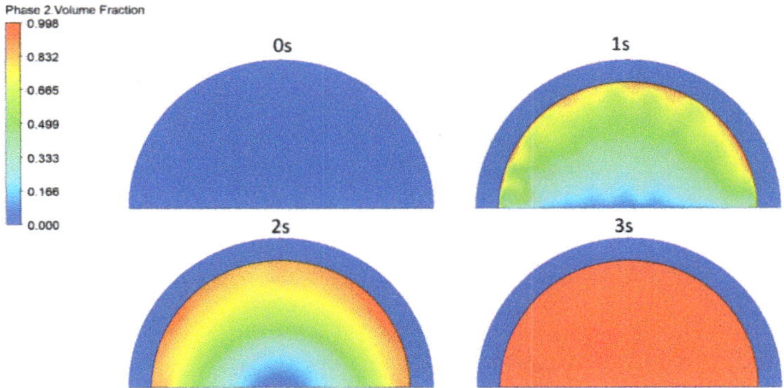

Figure 9. The simulated gas phase change during the period of 3 s.

Figure 10 presents the relation between temporal evolution of pressure inside the cancerous blister and the volume of fluid which has not yet undergone phase change (evaporated). The maximum calculated pressure is 817 kPa, after 3 s of laser irradiation. At that moment, 96% of the liquid had already evaporated. At that moment, the pressure forces surpass the elastic forces and explosion results.

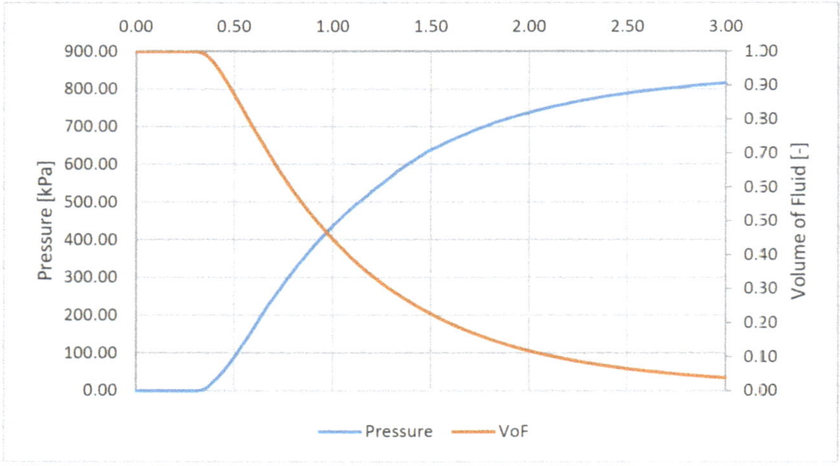

Figure 10. Relation between volume of fluid inside skin blister and pressure inside.

4. Discussions

Diode lasers with wavelengths of 810 ± 10 and 980 ± 10 nm are used in cw and pulsed modes. The radiation is transmitted to the operation field using optical fibre, which may or may not contact the tissue being treated. These DL's promote less bleeding, cleaner and more adequate operative field, significant reduction in post-operative oedema associated with direct reduction in pain, and improvement in tissue repair (see, e.g., [45]). Besides, Hanke et al. [46] studied soft-tissue cutting-efficiency for DL emitting in the wavelengths (λ) range 400 to 1500 nm. They found that the cutting depth for 2.5 W laser radiation moving at the speed 2 mm/s is equal 530, 330, 260, 230 μ for λ = 445, 810, 980, 1064 nm, respectively. Total interaction zones change accordingly. The 980 nm radiation is slightly better absorbed by water than the 810 nm one, which results in a smaller interaction zone. For example, Goel et al. [47] stated "The diode laser 980 nm is usually preferred for DCR surgery as it provides a better ablation and narrower tissue area involvement versus 810 nm that creates better coagulation than the vaporization". Table 1 presents various medical applications of the mentioned lasers.

Table 1. Medical application of pulsed and cw diode laser with wavelengths 810 and 980 nm (λ denotes wavelengths, τ pulse lengths).

λ (nm)	Operation Mode	Applications Field	Ref.
	cw and pulsed	photobiomodulation (PBM)—also known as biostimulation or low level laser therapy—LLLT, dental biostimulation, neuronal differentiation	[4,48]
	τ ~ 5–15 ms ok. 30 Hz	lesion tissue: telangiectasias, spider veins, capillary dermal malformation and a cutaneous venous malformation	[10]
810 ± 10	NA	vascular ophtalmology,	[49]
	cw and pulsed	congenital subglottic hemangiomas	[13]
	cw	Tis and T1 glottic carcinomas	[14]
	cw, 3.5 W	Pyogenic granuloma	[16]
	8–10 W τ~ 0.5–1 s	Laser assisted dacryocystorhinostomy	[43]
	p. 30 ms, 7 Hz	hair removal	[2[illegible]]
	5 W, τ = 4 s	Laryngology: laser-assisted somnoplasty, frenuloplasty, closure of tonsillar crypts, haemostasis, removal of cancerous changes, blepharoplasty, surgery of laryngopharynx and larynx, separation of adhesions in nasal septum	here

Table 1. *Cont.*

λ (nm)	Operation Mode	Applications Field	Ref.
980 ± 10	cw, pulsed	photobiomodulation (PBM)—also known as biostimulation or low level laser therapy—LLLT, wound and ulcers healing applications	[7]
	pulsed, cw	tissue tumors, frenectomies, excision of gingival hyperplasias, vestibuloplasties, hemangioma removal, and periimplant	[11]
	NA	turbinate reduction, nasal polypectomy, ablation of an oral papilloma, photocoagulation of nasal telangiectasias,	[23]
	NA	endovenous laser treatment of saphenous veins	[24]
	NA	telangiectasias	[12]
	cw	early glottic cancer	[27]
	cw	vocal fold polyps	[28]
	Ceralas D15, 12 W superpulse	laryngeal lesions: vocal polyps, leucoplakia, laryngeal hair, granuloma, subglottic stenosis and tracheal lesions	[29]
	cw, τ = 50 ms, 10 Hz	dermatology: neurofibroma, hemangioma	[34]
	dual 980/1470 nm	vascular lesions of skin and lips: cherry angiomas, venous lakes, lip hemangioma, and spider nevi, couperose, facial telangiectasia	[50]
	cw, 30–120 W τ = 0.1 s, 5–9 Hz	prostate treatment	[51]
	NA	ophthalmology	[47]
	3 W (100 Hz)	removal of intracanal *Enterococcus faecalis* biofilm	[32]
	cw 8 to 9 W	bilateral vocal fold immobility (BVFI)	[30,31]
	cw 3–4 W	Maxillofacial surgeries including intrinsic TMJ pathologies	[45]
	cw 1.5 W	gingival depigmentation	[52]

Although in the paper we focus on diode laser application, in otolaryngology various lasers have been used, following the first (in the late 70s) implementation of an argon laser for inferior turbinate reduction. Lasers have been later successfully applied for a variety of nasal pathologies, such as epistaxis, inferior turbinate hypertrophy, nasal and paranasal tumors, skin lesions, and pathologies of the nasopharynx—see e.g., [53]. Although, Abiri et al. [54] pointed to the argon laser as the superior for some laryngology problems (caused, e.g., by hereditary haemorrhagic telangiectasia) other lasers such as Nd:YAG (second harmonic) and diode lasers also give good results. However, the application of CO_2 laser radiation is limited due to the complexity of nasal anatomy and lack of appropriate elastic fibres.

The CO_2, Nd:YAG (second harmonic), argon and diode lasers were also successfully applied to oral cavity and oropharyngeal lesions, such as hypertrophic gingivitis, chronic tonsillitis, benign and malignant tumors, etc. [55]. These lasers provide better haemostasis, greater cutting precision, and reduced postoperative edema when compared to other standard methods of surgery.

The first laser assisted dacryocystorhinostomy (DCR) was implemented (by Massaro et al. [56]) in order to increase the diameter of tear duct (nasolacrimal duct) whilst avoiding bleeding. The argon laser was used in order to generate a tear duct (4–6 mm wide), which allow tears from the lacrimal sac of the eye to reach the nasal cavity. Later, the advantage of various elastic fibres allowed the application of other wavelengths, e.g., 2120 nm of Ho:YAG laser [57], 810 [43] or 980 nm [58] diode lasers.

Fluence is a key parameter which should be carefully adjusted in order to cause minimal damage to tissues adjacent to the incision site. Another issue is related to pulse operation mode. It was observed during neurofibroma treatment that pulsed mode (50 ms, 10 Hz) was perceived by the patient as less painful than the cw regime. However, it led to a longer operation time. Besides, using a higher pulse power for a shorter period of time results in less tissue damage than using lower power for a longer period of time. This is of special importance for selective photothermolises studied by Anderson and Parrish [3], but the most popular diode laser used in medical therapies does not offer such possibilities.

Therefore, cw mode operation is usually favoured in various therapies due to the reduced operation time. The exceptions are presented in Table 1.

5. Conclusions

The results of radiation tissue interactions depend upon the tissue absorption coefficient, the wavelength of the laser, power density, operation mode (including pulse lengths and frequency), and interaction time. Although these data are presently better described in various papers they are still not always fully provided.

From Table 1, it is clear that in the case of soft tissue surgery the cw operation mode is preferred by the medical staff. This is because of the limitations of pulse power in the most common diode lasers and its effect on operation time. However, one should remember that pulsed operation mode may result in less damage in tissue adjacent to the incision site. The 980 nm DL radiation may in some cases provide a better ablation and narrower tissue affected zone in relation to 810 nm laser which in turn will be better for coagulation.

Diode lasers are becoming increasingly popular in medical applications due to their small size, robustness and compactness, cost-effectiveness, and ease of operation as well as high efficiency (reaching up to 70%). Moreover, the price of diode lasers is getting more and more competitive in relation to other systems. However, the significant drawback of this technology is the scarcity of diode lasers with short and high power pulses, important, e.g., in the case of selective photothermolises therapy [3]. Pulse powers up to 150 W are available [59].

Author Contributions: Conceptualization, M.M., T.O and A.C.; methodology, M.M., J.S. and T.O; software, M.S.; validation, M.M., T.O. and A.C.; formal analysis, M.M., J.S., T.O. and A.C.; investigation, M.M., J.S. and M.S.; resources, M.M., T.O. and A.C.; data curation, M.M. and M.S.; writing—original draft preparation, M.M., M.S., T.O. and A.C.; writing—review and editing, M.M., J.S. and A.C.; visualization, M.M, J.S. and M.S.; supervision, M.M., T.O. and A.C.; project administration, M.M. and A.C.; funding acquisition, M.M and A.C. All authors have read and agreed to the published version of the manuscript.

Funding: This research received no external funding.

Informed Consent Statement: Informed consent was obtained from all subjects involved in the study as mentioned in Appendix A.

Conflicts of Interest: There are no conflicts of interest.

Appendix A. Patient Statement of Informed Consent for Surgical Treatment

As a patient of MML Medical Center, I hereby agree to undergo treatment, which will consist of:

(Name of procedure)

I have been informed of the requirements, processes and stages of surgical treatment, its purpose, expected results and potential risks that may occur as a result of this treatment. I accept the multi-disciplinary treatment plan, which involves surgical treatment. I have been informed of the costs of treatment and accept these.

I have been informed of the possibility of early and late post-surgery complications and the accompanying risks. I have understood the explanations and asked all questions that are of interest to me in regard to this medical procedure. Should a situation arise requiring it, I agree to a modification of the surgical procedure to the necessary extent, in accordance with the principles of medical knowledge.

I hereby give conscious consent to perform this treatment under local/general anaesthesia and declare that I have not concealed any crucial information regarding my overall health status. I have been informed about the possibilities of medical complications during the procedure which will be conducted.

I have been informed of and agree to allow the necessary photographic and radiological documentation in connection with the treatments.

I have been informed about the necessity of reporting to post-treatment follow-up control visits. I have been informed that smoking and poor oral hygiene and failure to follow post-treatment recommendations can significantly exacerbate potential post-treatment complications and negatively affect the treatments success.

I submit to the following restrictions associated with the medical procedure:

Performing Doctor Legible Patient signature

References

1. Goldman, L.; Blaney, D.J.; Kindel, D.J.; Franke, E.K. Effect of the Laser Beam on the Skin**From the Departments of Dermatology and Physics of the University of Cincinnati. The study was done under a grant from the U.S. Public Health Service 0H0018. *J. Investig. Dermatol.* **1963**, *40*, 121–122. [CrossRef]
2. Song, K.U. Footprints in Laser Medicine and Surgery: Beginnings, Present, and Future. *Med. Lasers* **2017**, *6*, 1–4. [CrossRef]
3. Anderson, R.; Parrish, J. Selective photothermolysis: Precise microsurgery by selective absorption of pulsed radiation. *Science* **1983**, *220*, 524–527. [CrossRef]
4. Mester, A. Laser Biostimulation. *Photomed. Laser Surg.* **2013**, *31*, 237–239. [CrossRef]
5. Hamblin, M.R. Photobiomodulation or low-level laser therapy. *J. Biophotonics* **2016**, *9*, 1122–1124. [CrossRef]
6. Mester, E.; Szende, B.; Spiry, T.; Scher, A. Stimulation of wound healing by laser rays. *Acta Chir. Acad. Sci. Hung.* **1972**, *13*, 315–324.
7. Kawalec, J.; Reyes, C.; Penfield, V.; Hetherington, V.; Hays, D.; Feliciano, F.; Gartz, D.; Jones, J.; Esposito, R.; Cernica, M. Evaluation of the Ceralas D15 diode laser as an adjunct tool for wound care: A pilot study. *Foot* **2001**, *11*, 68–73. [CrossRef]
8. Goo, H.; Kim, H.; Ahn, J.-C.; Cho, K.J. Effects of Low-level Light Therapy at 740 nm on Dry Eye Disease In Vivo. *Med. Lasers* **2019**, *8*, 50–58. [CrossRef]
9. Espey, B.T.; Kielwein, K.; van der Ven, H.; Steger, K.; Allam, J.; Paradowska-Dogan, A.; van der Ven, K. Effects of Pulsed-Wave Photobiomodulation Therapy on Human Spermatozoa. *Lasers Surg. Med.* **2021**. [CrossRef]
10. Bass, L.S. Photosclerosis of cutaneous vascular malformations with a pulsed 810-nm diode laser. In *Lasers in Surgery: Advanced Characterization, Therapeutics, and Systems V, Proceedings of the Photonics West '95, San Jose, CA, USA, 1−28 February 1995*; SPIE: Bellingham, WA, USA, 1995; pp. 559–565. [CrossRef]
11. Romanos, G.; Nentwig, G.-H. Diode Laser (980 nm) in Oral and Maxillofacial Surgical Procedures: Clinical Observations Based on Clinical Applications. *J. Clin. Laser Med. Surg.* **1999**, *17*, 193–197. [CrossRef] [PubMed]
12. Desiate, A.; Cantore, S.; Tullo, D.; Profeta, G.; Grassi, F.R.; Ballini, A. 980 nm diode lasers in oral and facial practice: Current state of the science and art. *Int. J. Med. Sci.* **2009**, *6*, 358–364. [CrossRef]
13. Saetti, R.; Silvestrini, M.; Cutrone, C.; Narne, S. Treatment of Congenital Subglottic Hemangiomas. *Arch. Otolaryngol. Head Neck Surg.* **2008**, *134*, 848–851. [CrossRef] [PubMed]
14. Ferri, E.; Armato, E. Diode laser microsurgery for treatment of Tis and T1 glottic carcinomas. *Am. J. Otolaryngol.* **2008**, *29*, 101–105. [CrossRef] [PubMed]
15. Mittnacht, D.; Linder, A.; Foth, H.-J. Medical Application of a High Power Diode Laser. In Therapeutic Laser Applications and Laser-Tissue Interactions II, In Proceedings of the SPIE—The International Society for Optical Engineering, Munich, Germany, 12–16 Jun 2005; Optical Society of America: Washington, DC, USA, 2005.
16. Azma, E.; Safavi, N. Diode Laser Application in Soft Tissue Oral Surgery. *J. Lasers Med. Sci.* **2013**, *4*, 206–211. [PubMed]
17. Lee, D.Y.; Cho, J.-G.; Im, N.-R.; Lee, H.-J.; Kim, B.; Jung, K.-Y.; Kim, T.H.; Baek, A.S.-K. Evaluation of the Efficacy of 1940-nm Diode Laser in Tonsillectomy: Preliminary Report. *Med. Lasers* **2015**, *4*, 29–34. [CrossRef]
18. Kang, S.H.; Lim, S.; Oh, D.; Kang, K.; Jung, K.J.; Kim, H.K.; Lee, S.H.; Baek, S.-K.; Kim, A.T.H. Clinical Feasibility Trial of 1,940-nm Diode Laser in Korean Patients with Inferior Turbinate Hypertrophy. *Med. Lasers* **2015**, *4*, 60–64. [CrossRef]
19. Agrawal, A.A.; Prabhu, R.; Sankhe, R.; Wagle, S.V. Influence of external chromophore on cutting efficacy of 940 nm diode laser: An In vitro animal tissue study. *Contemp. Clin. Dent.* **2018**, *9*, 79–S82. [CrossRef]
20. Im, N.-R.; Kim, B.; Kim, J.; Baek, S.-K. Treating Oral Leukoplakia with a 532-nm Pulsed Diode Laser. *Med. Lasers* **2019**, *8*, 39–42. [CrossRef]
21. Kang, H.Y.; Park, E.S.; Nam, S.M. Prospective, Comparative Evaluation of Forearm and Lower Leg Hair Removal with 808-nm Diode Laser at Different Fluences. *Med. Lasers* **2018**, *7*, 55–61. [CrossRef]
22. Im, N.-R.; Moon, J.; Choi, W.; Kim, B.; Lee, J.J.; Kim, H.; Baek, S.-K. Histologic Evaluation of Blood Vessels Sealed with 1,470-nm Diode Laser: Determination of Adequate Condition for Laser Vessel Sealing. *Med. Lasers* **2018**, *7*, 6–12. [CrossRef]
23. Newman, J.; Anand, V. Applications of the diode laser in otolaryngology. *Ear Nose Throat J.* **2002**, *81*, 850–851. [CrossRef]
24. Schmedt, C.-G.; Sroka, R.; Steckmeier, S.; Meissner, O.A.; Babaryka, G.; Hunger, K.; Ruppert, V.; Sadeghi-Azandaryani, M.; Steckmeier, B.M. Investigation on Radiofrequency and Laser (980 nm) Effects after Endoluminal Treatment of Saphenous Vein Insufficiency in an Ex-vivo Model. *Eur. J. Vasc. Endovasc. Surg.* **2006**, *32*, 318–325. [CrossRef]
25. Reynaud, J.P.; Skibinski, M.; Wassmer, B.; Rochon, P.; Mordon, S. Lipolysis using a 980-nm diode laser: A retrospective analysis of 534 procedures. *Aesthetic Plast. Surg.* **2009**, *33*, 28–36. [CrossRef]
26. Weiss, R.A.; Beasley, K. Laser-assisted liposuction using a novel blend of lipid- and water-selective wavelengths. *Lasers Surg. Med.* **2009**, *41*, 760–766. [CrossRef] [PubMed]

27. Tunçel, Ü.; Cömert, E. Preliminary Results of Diode Laser Surgery for Early Glottic Cancer. *Otolaryngol. Neck Surg.* **2013**, *149*, 445–450. [CrossRef] [PubMed]
28. Karasu, M.F.; Gundogdu, R.; Cagli, S.; Aydin, M.; Arli, T.; Aydemir, S.; Yüce, I.; Aydın, M. Comparison of Effects on Voice of Diode Laser and Cold Knife Microlaryngology Techniques for Vocal Fold Polyps. *J. Voice* **2014**, *28*, 387–392. [CrossRef] [PubMed]
29. Hwang, S.M.; Lee, D.Y.; Im, N.-R.; Lee, H.-J.; Kim, B.; Jung, K.-Y.; Kim, T.H.; Baek, A.S.-K. Office-Based Laser Surgery for Benign Laryngeal Lesion. *Med. Lasers* **2015**, *4*, 65–69. [CrossRef]
30. Karkos, P.D.; Stavrakas, M. Minimizing revision rates with the "Π" technique for bilateral vocal fold immobility: A new technique combining carbon dioxide and diode laser. *Head Neck* **2015**, *38*, 801–803. [CrossRef]
31. Karkos, P.D.; Stavrakas, M.; Koskinas, I.; Markou, K.; Triaridis, S.; Constantinidis, J. 5 Years of Diode Laser "Π" Technique for Bilateral Vocal Fold Immobility: A Technique That Improves Airway and Is Friendly to the Voice. *Ear Nose Throat J.* **2021**, *100*, 83S–86S. [CrossRef]
32. Prażmo, E.; Godlewska, A.; Sałkiewicz, M.; Mielczarek, A. Effects of 980 nm diode laser application protocols on the reduction of Enterococcus faecalis intracanal biofilm: An in vitro study. *Dent. Med. Probl.* **2017**, *54*, 333–338. [CrossRef]
33. Wróbel, M.S.; Jędrzejewska-Szczerska, M.; Galla, S.; Piechowski, L.; Sawczak, M.; Popov, A.P.; Bykov, A.V.; Tuchin, V.V.; Cenian, A. Use of optical skin phantoms for preclinical evaluationof laser efficiency for skin lesion therapy. *J. Biomed. Opt.* **2015**, *20*, 085003. [CrossRef] [PubMed]
34. Szymańczyk, J.; Sawczak, M.; Cenian, W.; Karpienko, K.; Jędrzejewska-Szczerska, M.; Cenian, A. Application of the laser diode with central wavelength 975 nm for the therapy of neurofibroma and hemangiomas. *J. Biomed. Opt.* **2017**, *22*, 010502. [CrossRef]
35. Milanic, M.; Cenian, A.; Verdel, N.; Cenian, W.; Stergar, J.; Majaron, B. Temperature Depth Profiles Induced in Human Skin In Vivo Using Pulsed 975 nm Irradiation. *Lasers Surg. Med.* **2019**, *51*, 774–784. [CrossRef]
36. Piechowski, L.; Cenian, W.; Sawczak, M.; Cenian, A. Pulsed dermatologic 20W diode-laser emitting at 975 nm. In *Laser Technology 2012: Applications of Lasers, Proceedings of the Tenth Symposium on Laser Technology, Bellingham, WA, USA, 24−28 September 2012*; International Society for Optics and Photonics: Bellingham, WA, USA, 2012.
37. Bodnar, T.; Galdi, G.P.; Necasova, S. *Fluid-Structure Interaction and Biomedical Applications*; Springer: Basel, Switzerland, 2014.
38. Badur, J.; Ziółkowski, P.; Zakrzewski, W.; Sławiński, D.; Kornet, S.; Kowalczyk, T.; Hernet, J.; Piotrowski, R.; Felincjancik, J.; Ziółkowski, P.J. An advanced Thermal–FSI approach to flow heating/cooling. *J. Phys. Conf. Ser.* **2014**, *530*, 012039. [CrossRef]
39. Ziółkowski, P.J.; Ochrymiuk, T.; Eremeyev, V.A. Adaptation of the arbitrary Lagrange–Euler approach to fluid–solid interaction on an example of high velocity flow over thin platelet. *Contin. Mech. Thermodyn.* **2019**, 1–14. [CrossRef]
40. Maitland, D.J.; Eder, D.C.; London, R.A.; Glinsky, M.E.; Soltz, B.A. Dynamic simulations of tissue welding. *Proc. Soc. Photo-Opt. Ins.* **1996**, *2671*, 234–242.
41. Zhou, J.; Liu, J.; Yu, A. Numerical Study on the Thawing Process of Biological Tissue Induced by Laser Irradiation. *ASME. J. Biomech. Eng.* **2005**, *127*, 416–431.
42. Michalik, M.; Podbielska-Kubera, A.; Dmowska-Koroblewska, A. Diode laser-assisted uvulopalatoplasty using palisade technique. *New Med.* **2018**, *22*. [CrossRef]
43. Broda, M.; Michalik, M.; Białas, D.; Różycki, R. Laser dacryocystorhinostomy—The lasers use in treatment of lacrimal duct obstruction. *OphthaTherapy Ther. Ophthalmol.* **2018**, *5*, 109–113. [CrossRef]
44. Michalik, M. The influence of mitomycin C on clinical results of the nasolacrimal duct anastomosis using the diode laser. *OphthaTherapy Ther. Ophthalmol.* **2019**, *6*, 274–278. [CrossRef]
45. Guerra, R.C.; de Luca, D.N.; Pereira, R.S.; Carvalho, P.H.A.; Homsi, N.; Radaic, P.; Pastore, G.P.; Hochuli-Vieira, E. TMJ diode surgical laser approach in a contemporary treatment of tempomandibular joint pathologies. A technical note. *Oral Surg.* **2021**, *14*, 206–208. [CrossRef]
46. Hanke, A.; Fimmers, R.; Frentzen, M.; Meister, J. Quantitative determination of cut efficiency during soft tissue surgery using diode lasers in the wavelength range between 400 and 1500 nm. *Lasers Med. Sci.* **2021**, 1–15. [CrossRef]
47. Goel, R.; Nagpal, S.; Garg, S.; Malik, K.P.S. Is transcanalicular laser dacryocystorhinostomy using low energy 810 nm diode laser better than 980 nm diode laser? *Oman J. Ophthalmol.* **2015**, *8*, 134. [CrossRef]
48. George, S.; Hamblin, M.R.; Abrahamse, H. Photobiomodulation-Induced Differentiation of Immortalized Adipose Stem Cells to Neuronal Cells. *Lasers Surg. Med.* **2020**, *52*, 1032–1040. [CrossRef]
49. Akduman, L.; Olk, R.J. Diode Laser (810 nm) versus Argon Green (514 nm) Modified Grid Photocoagulati-on for Diffuse Diabetic Macular Edema. *Ophthalmology* **1997**, *104*, 1433–1441. [CrossRef]
50. Wollina, U.; Goldman, A. The dual 980-nm and 1470-nm diode laser for vascular lesions. *Dermatol. Ther.* **2020**, *33*, 085003. [CrossRef] [PubMed]
51. Wendt-Nordahl, G.; Huckele, S.; Honeck, P.; Alken, P.; Knoll, T.; Michel, M.S.; Häcker, A. 980-nm Diode Laser: A Novel Laser Technology for Vaporization of the Prostate. *Eur. Urol.* **2007**, *52*, 1723–1728. [CrossRef]
52. Arif, R.H.; Kareem, F.A.; Zardawi, F.M.; Al-Karadaghi, T.S. Efficacy of 980 nm diode laser and 2940 nm Er: YAG laser in gingival depigmentation: A comparative study. *J. Cosmet. Dermatol.* **2021**, *20*, 1684–1691. [CrossRef]
53. Betka, J.; Plzák, J.; Zábrodský, M.; Kastner, J.; Boucek, J. Lasers in otorhinolaryngology (ORL) and head and neck surgery. In *Lasers for Medical Applications*; Woodhead Publishing: Sawston, UK, 2013; pp. 556–572. [CrossRef]
54. Abiri, A.; Bs, K.G.; Maducdoc, M.; Sahyouni, R.; Wang, M.B.; Kuan, E.C. Laser-Assisted Control of Epistaxis in Hereditary Hemorrhagic Telangiectasia: A Systematic Review. *Lasers Surg. Med.* **2020**, *52*, 293–300. [CrossRef] [PubMed]

55. Ferlito, S.; Nane, S.; Grillo, C.; Maugeri, M.; Cocuzza, S.; Grillo, C. Diodes laser in ENT surgery. *Acta Med. Mediterr.* **2011**, *27*, 79.
56. Massaro, B.M.; Gonnering, R.S.; Harris, G.J. Endonasal laser dacryocystorhinostomy. A new approach to nasolacrimal duct obstruction. *Arch. Ophthalmol.* **1990**, *108*, 1172–1186. [CrossRef] [PubMed]
57. Silkiss, R.Z.; Axelrod, R.N.; Iwach, A.G.; Vassiliadis, A.; Hennings, D.R. Transcanalicular THC: YAG dacryocystorhinostomy. *Ophthalmic Surg.* **1992**, *23*, 351–353. [PubMed]
58. Nowak, R.; Rekas, M.; Gospodarowicz, I.N.; Ali, M.J. Long-term outcomes of primary transcanalicular laser dacryocystorhinostomy. *Graefe's Arch. Clin. Exp. Ophthalmol.* **2021**, *54*, 1–6. [CrossRef]
59. Yaroslavsky, I.; Boutoussov, D.; Vybornov, A.; Perchuk, I.; Meleshkevich, V.; Altshuler, G.B. Ex vivo evaluation of super pulse diode laser system with smart temperature feedback for contact soft-tissue surgery. In *Lasers in Dentistry XXIV, Proceedings of the SPIE BiOS, Bellingham, WA, USA, 27 January—1 February 2018*; International Society for Optics and Photonics: Bellingham, WA, USA, 2018; p. 104730. [CrossRef]

MDPI

Article

Microstructure Study of Pulsed Laser Beam Welded Oxide Dispersion-Strengthened (ODS) Eurofer Steel

Jia Fu [1,2,*], Ian Richardson [1] and Marcel Hermans [1]

1 Department of Materials Science and Engineering, Delft University of Technology, 2628 CD Delft, The Netherlands; I.M.Richardson@tudelft.nl (I.R.); M.J.M.Hermans@tudelft.nl (M.H.)
2 Dutch Institute for Fundamental Energy Research (DIFFER), 5600 HH Eindhoven, The Netherlands
* Correspondence: j.fu@tudelft.nl

Abstract: Oxide dispersion-strengthened (ODS) Eurofer steel was laser welded using a short pulse duration and a designed pattern to minimise local heat accumulation. With a laser power of 2500 W and a duration of more than 3 ms, a full penetration can be obtained in a 1 mm thick plate. Material loss was observed in the fusion zone due to metal vaporisation, which can be fully compensated by the use of filler material. The solidified fusion zone consists of an elongated dual phase microstructure with a bimodal grain size distribution. Nano-oxide particles were found to be dispersed in the steel. Electron backscattered diffraction (EBSD) analysis shows that the microstructure of the heat-treated joint is recovered with substantially unaltered grain size and lower misorientations in different regions. The experimental results indicate that joints with fine grains and dispersed nano-oxide particles can be achieved via pulsed laser beam welding using filler material and post heat treatment.

Keywords: oxide dispersion strengthened steel; ODS Eurofer; laser welding; microstructure; EBSD

Citation: Fu, J.; Richardson, I.; Hermans, M. Microstructure Study of Pulsed Laser Beam Welded Oxide Dispersion-Strengthened (ODS) Eurofer Steel. *Micromachines* **2021**, *12*, 629. https://doi.org/10.3390/mi12060629

Academic Editors: Congyi Wu and Yu Huang

Received: 29 April 2021
Accepted: 26 May 2021
Published: 28 May 2021

Publisher's Note: MDPI stays neutral with regard to jurisdictional claims in published maps and institutional affiliations.

1. Introduction

Due to their good high-temperature strength, corrosion resistance and radiation resistance, oxide dispersion-strengthened (ODS) steels are promising candidates for structural materials employed in elevated-temperature and nuclear applications [1]. The favourable properties of ODS steels are mainly attributed to the fine grains and homogenously dispersed nanosized oxide particles in the steel matrix [2]. These fine and thermally stable dispersoids hinder the motion of dislocations and grain boundaries, acting as trapping sites for both point defects and helium atoms generated during irradiation, resulting in an increased resistance to irradiation damage.

Despite the promising behaviour of ODS steels for use in advanced nuclear systems, joining these materials remains one of the major technological challenges limiting their deployment [3]. Joining ODS steels by solid-state methods such as spark plasma sintering (SPS), hot isostatic pressing (HIP) and friction stir welding (FSW) has been proven to be feasible by several authors [4–6]. The degradation of featured microstructures and mechanical properties can be minimised since these techniques do not create a molten zone in the joint area [5]. However, the costs of SPS and HIP are relatively high due to long processing times (1–5 h) [7] and the application of FSW is limited due to geometrical restrictions and tool wear [8]. The welding of ODS steels by traditional, fusion-based welding techniques such as gas metal arc welding and tungsten inert gas welding is problematic. As soon as a molten zone is produced, the oxide particles rapidly agglomerate and float to the top of the molten weld pool, resulting in a significant loss of strength [9]. Laser beam welding [10–12] can potentially be employed for joining ODS steels due to its highly concentrated energy input, leading to the melting of a small amount of base material, and consequently, the formation of a small heat-affected zone (HAZ). The study of Lemmen et al. [12] showed that PM1000 had a good laser weldability with a wide range of welding parameters. However, yttrium oxide clustering was found in all conditions, causing a reduction in strength in the weld. Similar results were obtained by Liang et al. [13] who indicated that the nanoprecipitates were larger in the weld metal than in the base material. In summary, a new laser welding method needs to be developed to address the issue of microstructure and mechanical behaviour degradation.

In this study, pulsed laser beam welding was successfully employed to join ODS Eurofer steel with only minor deterioration of microstructure when compared to the parent material. The welding parameters were investigated and optimised to improve the microstructure of the joint. The microstructural features were characterised by means of optical microscopy (OM), scanning electron microscopy (SEM), transmission electron microscopy (TEM) and electron backscattered diffraction (EBSD).

2. Experimental Details

2.1. Materials

The alloy studied was ODS Eurofer steel with a nominal composition of Fe–9Cr–1.1W–0.4Mn–0.2V–0.12Ta–0.1C–0.3Y_2O_3 (wt%), produced via powder metallurgy. The production process started with mechanical alloying, where the precursor powders were mixed in a Retsch planetary ball milling machine under an argon atmosphere for 30 h at 300 rpm. The resultant powders were subsequently consolidated by spark plasma sintering (SPS, FCT group, Frankenblick, Germany) at a pressure of 60 MPa and a sintering temperature of 1373 K with a heating rate of 100 K/min. After a holding time of 30 min, disks of 40 mm diameter and around 10 mm thickness were produced (Figure 1). These parameters were selected based on our previous study [14].

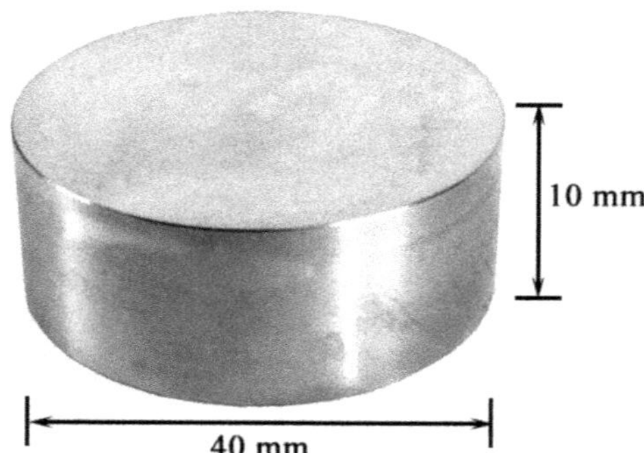

Figure 1. Powder metallurgy-produced ODS Eurofer.

2.2. Methods

A 6 kW Yb:YAG laser was used for the welding experiments. The focusing optic has a focal length of 223 mm and projects a laser spot with a diameter of 0.6 mm. Two work pieces (30 × 20 × 1 mm^3) machined from the SPS-prepared disk were welded by pulsed laser beam welding with a peak power of 2500 W and a pulse duration ranging from 2 ms to 5 ms. A shielding gas of argon was delivered to the work piece at a flow rate of 8 L/min. Instead of moving straight in one direction, the laser beam was moved following the sequence indicated in Figure 2 in order to minimise the heat accumulation in the material and, therefore, shorten the melt pool lifetime. The time interval between each point is around 30 s. The distance between the centres of adjacent spots was 0.5 mm to ensure a continuous weld. A post heat treatment was conducted to recover the microstructure and release the residual stress generated during welding by normalising at 1423 K for 1 h, air cooling to room temperature, and then, tempering at 973 K for 1 h, followed by air cooling to room temperature.

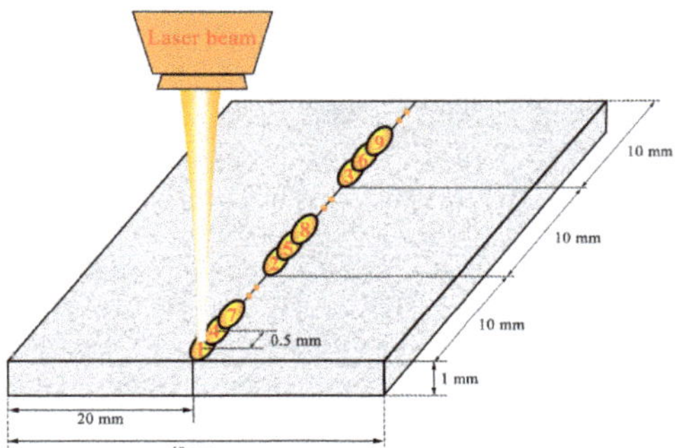

Figure 2. Schematic illustration of welding strategies.

The microstructure of the joint was characterised using a Keyence Digital Microscope VHX-5000 and a JEOL 6500F SEM equipped with an energy dispersive spectrometer (EDS) and EBSD. The nano-oxide particles in the material were investigated using a JEM-2200FS TEM. To reveal the microstructure by OM and SEM, the samples were etched in a solution of 5 g ferric chloride, 50 mL HCl and 100 mL distilled water for 20 s. Specimens for EBSD were mirror polished followed by a colloidal silica polishing step. TSL orientation imaging microscopy (OIM) software was used for data processing and analysis. EBSD maps of inverse pole figure (IPF), grain average image quality (GAIQ) and kernel average misorientation (KAM) were implemented and analysed in the study. The TEM specimens were prepared by electropolishing disks with a diameter of 3 mm in a twin-jet electropolisher using 4% perchloric acid and 96% ethanol as electrolyte.

3. Results

3.1. Parameter Optimisation

To study the effect of a pulsed laser beam on the microstructure of ODS Eurofer, a number of spots were created on a plate with varying parameters. A short melt pool lifetime would be beneficial for retaining the Y_2O_3 particles in the fusion zone. Therefore, a laser power of 2500 W and short pulse durations between 2 ms and 4 ms were applied. Optical micrographs of the cross-section of the spots can be seen in Figure 3. Material loss was observed in all conditions due to metal evaporation during the welding process. The width of the top of the "V"-shaped fusion zone is around 0.7 mm, which is very close to the beam size, indicating a concentrated heat input. The heat-affected zone (HAZ) is small in all cases, with a width of approximately 0.06 mm. It can be seen that partial penetration is obtained with a pulse duration of 2 ms and 2.5 ms. Large pores can be observed in the bottom of the fusion zone, probably because gas was trapped in the melt pool due to a short escape time. Full penetration is realised with pulse durations of more than 2.5 ms. As expected, more severe material loss was observed with longer laser beam pulses. Large pores managed to escape in these open keyhole conditions, while microvoids were found in the weld pool.

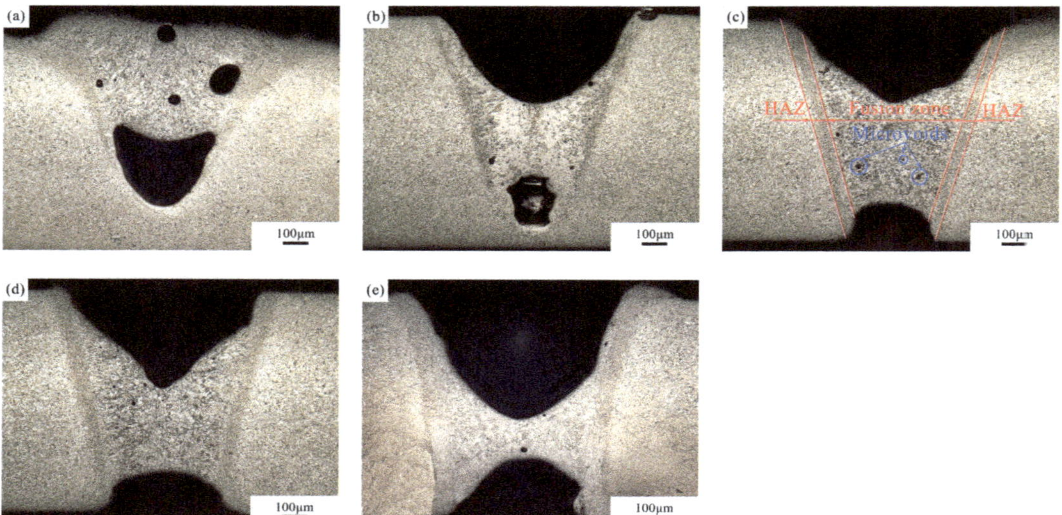

Figure 3. Optical micrographs of spots on the plate, with a power of 2500 W and a pulse duration of (**a**) 2 ms, (**b**) 2.5 ms, (**c**) 3 ms, (**d**) 3.5 ms and (**e**) 4 ms.

3.2. Microstructure Characterisation

In order to compensate for the material loss in the fusion zone, two ODS Eurofer square bars with dimensions of 30 × 1 × 0.5 mm^3 were attached to the top and bottom surfaces of the work piece to act as a filler material. A duration of 5 ms was used to realise full penetration. The material was joined using the pattern indicated in Figure 2. An SEM image of the weld seam is shown in Figure 4a. It can be seen that the material loss in the specimen is fully compensated by the filler material. In Figure 4b, the microstructure has both martensite grains (dark regions) and ferrite grains (bright regions), with the presence of microvoids. Martensite is formed during the rapid cooling process (~10^4–10^6 K s^{-1} [15]), while δ-ferrite is formed during heating, as the peak temperature of laser welding is definitely higher than the austenite–δ-ferrite transformation temperature (around 1400 K). However, since the δ-ferrite–austenite transformation is a diffusion-controlled process, the rapid solidification does not offer sufficient time to complete the phase transformation [16]. Consequently, residual ferrite is observed in the microstructure.

Figure 4d shows an enlarged image of the microstructure in the fusion zone. A large number of nanoprecipitates can be observed in the steel matrix. As shown in the TEM images in Figure 5a,b, finely dispersed Y_2O_3 nanoparticles are observed in the microstructure. The particle sizes vary between 1 and 30 nm and do not show a significant difference in distribution between the fusion zone and base material. Figure 5c shows a dark field image of Y_2O_3 particles (indicated by the arrows) pinning the grain boundaries in the fusion zone, which is beneficial for enhancing the mechanical properties and extending the working temperature range. The martensite lath structure in the fusion zone is revealed in Figure 5d. From Figure 4d, it is also worth noting that the Y_2O_3 particles are not homogeneously distributed in the steel matrix. It seems that the smaller grains have a higher number of Y_2O_3 precipitates than the larger grains. This can be explained as follows: The distribution of Y_2O_3 is not perfectly homogenous even after a long period of mechanical alloying. Since Y_2O_3 nanoprecipitates have a strong effect on impeding grain growth through a Zener-type pinning [17], grains with a higher density of Y_2O_3 are presumably more resistant to recovery and growth.

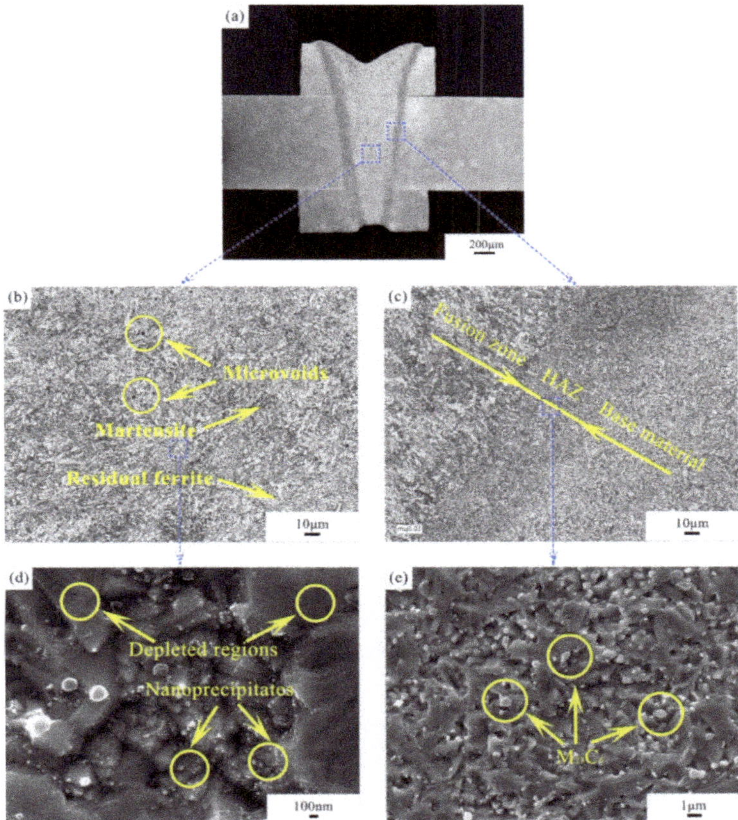

Figure 4. Morphologies of the weld seam with filler material: (**a**) SEM morphology of the weld seam, (**b**) fusion zone, (**c**) fusion line, (**d**) enlarged fusion zone and (**e**) enlarged HAZ.

The grain size of the HAZ is smaller compared to that of the base material and clearly smaller than that of the fusion zone (Figure 4c), probably due to martensite transformation. An enlarged image of the HAZ (Figure 4e) shows a large number of precipitates with a size ranging from 0.1 to 1 μm in the steel matrix. They are found to be more preferentially located at the grain boundaries, possibly due to a decrease in volume-free energy [17]. These precipitates are rich in Fe, Cr, W and C, which can be identified as $M_{23}C_6$ carbides (M = Fe, Cr and W) based on their size and chemical composition.

Figure 6 shows inverse pole figure (IPF) maps of the base material as well as the centre of the fusion zone and HAZ (P = 2500 W, t = 5 ms, with filler material) obtained by EBSD. Ultrafine grains smaller than 250 nm could not be indexed, causing the non-indexed (black) areas in the figures. It can be observed that the maps exhibit grain sizes ranging from the nanometre scale to a few micrometres. The microstructure of the fusion zone consists of elongated structures, while that of HAZ shows equiaxed grains. None of the regions show preferential grain orientation.

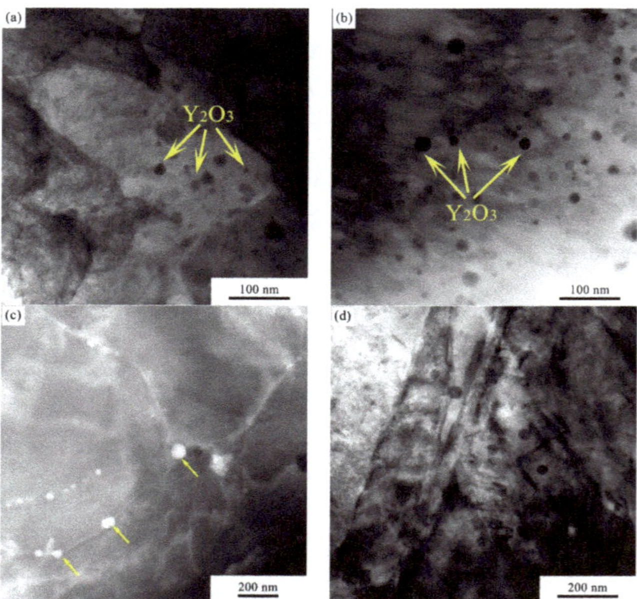

Figure 5. TEM images of the Y_2O_3 nanoparticles in the (**a**) base material, (**b**) fusion zone, (**c**) Y_2O_3 pinning the grain boundaries and (**d**) martensite lath.

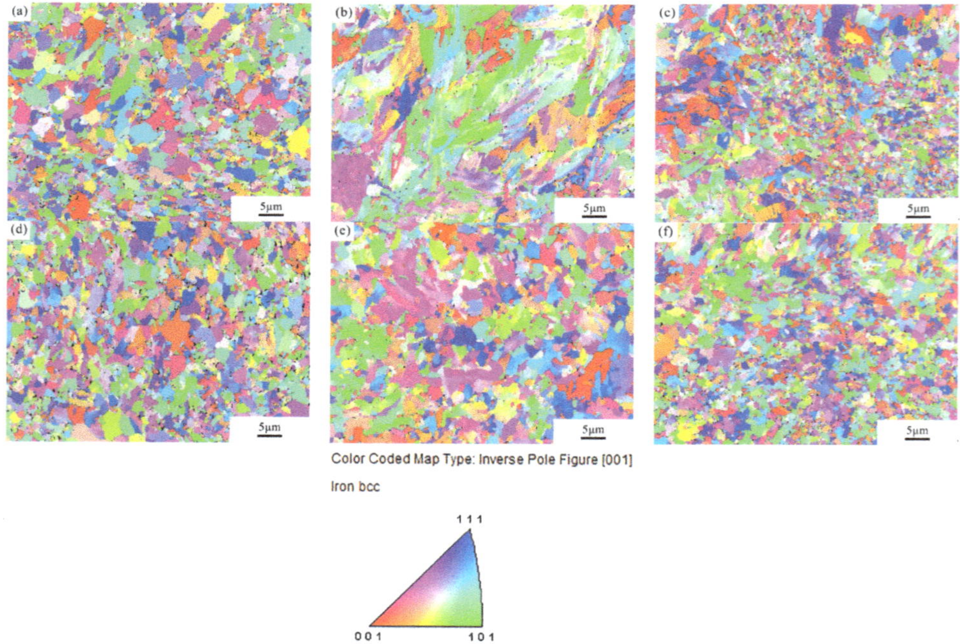

Figure 6. Inverse pole figure (IPF) maps of different regions: (**a**) base material, (**b**) fusion zone, (**c**) HAZ, (**d**) heat-treated base material, (**e**) heat-treated fusion zone and (**f**) heat-treated HAZ.

The average grain sizes were measured and are shown in Table 1. The grain size distribution was determined under the assumption that the minimum misorientation characterising grain boundaries is 15°. It can be seen that in the as-joined condition, the average grain size of different regions decreases from the fusion zone to the base material to the HAZ, which agrees with the observation from SEM. In the heat-treated condition, the average grain size of the fusion zone is still the largest, and that of the HAZ remains the smallest, but with smaller differences compared to the as-joined condition. Additionally, compared to the as-joined condition, there is a decrease in the grain size of the fusion zone, which could be ascribed to a refinement effect due to martensite transformation. Grain growth occurs in the HAZ, presumably resulting from the release of high stored energy due to phase transformation and distortion during laser welding. In general, the grain sizes of the joint overall do not grow significantly after the heat treatment, probably because of the strong pinning effect of Y_2O_3 on the motion of grain boundaries.

Table 1. Average grain size of different regions obtained from EBSD data in Figure 6.

Conditions	Regions	Grain Size/μm
As-joined	Base material	2.08 ± 1.31
	Fusion zone	4.33 ± 2.89
	HAZ	1.64 ± 1.25
Heat treated	Base material	1.92 ± 1.10
	Fusion zone	2.20 ± 1.50
	HAZ	1.72 ± 1.07

From Table 1, it can also be noted that the standard deviations of all conditions are high. This is due to a bimodal grain size distribution of the material. Taking the fusion zone in the heat-treated joint (Table 1) as an example, the grain size distribution has two peaks at 1.42 and 7.20 μm, respectively, as can be seen in Figure 7. Similar bimodal grain size distributions have been widely reported for powder metallurgy-prepared ODS steels [18–20]. In martensitic–ferritic steels, which generally contain 0.1–0.2 wt% C and 9–11 wt% Cr [21], this phenomenon could be due to the dual phase nature of the material. To further differentiate martensite from ferrite in the studied material, a chart of grain average image quality is shown in Figure 8a. Image quality (IQ) describes the quality of diffraction property of the analysed Kikuchi patterns. This factor can be used as an estimation of dislocation density or stored energy [22]. It can be seen that if a threshold value of approximately 40 is chosen, phases with image quality less than 40 are highlighted and shown in Figure 8b. In this way, martensite is differentiated from ferrite in the microstructure, since martensite exhibits lower image quality than ferrite due to its highly distorted lattice [23]. It can be noted that the smaller martensite grains are mainly located between the larger ferrite grains. This finding supports the dual phase microstructure of the material, and consequently, a high standard deviation of the grain size distribution.

The kernel average misorientation (KAM), which represents the numerical misorientation average of a given point with its nearest neighbours, was used to characterise local misorientations. The maps were calculated using a maximum misorientation of 5°. Figure 9a shows the KAM map of the fusion zone in the heat-treated joint. It can be observed that the grains standing out from the rest due to a larger size show almost no misorientation, i.e., no local lattice distortion. In addition, misorientations between 0.5° and 2° are observed near the grain boundaries. These small grain regions with higher misorientations can be linked to the areas with lower image quality. As shown in Figure 9b, the KAM map combined with the IQ map reveals overlapping regions, where martensite grains (yellow) have a higher misorientation and ferrite grains (dark blue) generally have a lower misorientation. By comparing KAM of different regions in the joint, this indicates whether the microstructure is changed significantly after welding or the heat treatment. The fusion zone and HAZ have a higher KAM than the base material, indicating a more distorted microstructure and probably a higher martensite fraction due to an additional thermal cycle

(Figure 10). Meanwhile, in the heat-treated condition, the KAM is smaller than that of the as-joined condition and shows no significant difference in different regions, demonstrating that the microstructure is substantially recovered and homogenous after the normalising and tempering treatment, which would be beneficial for the mechanical properties.

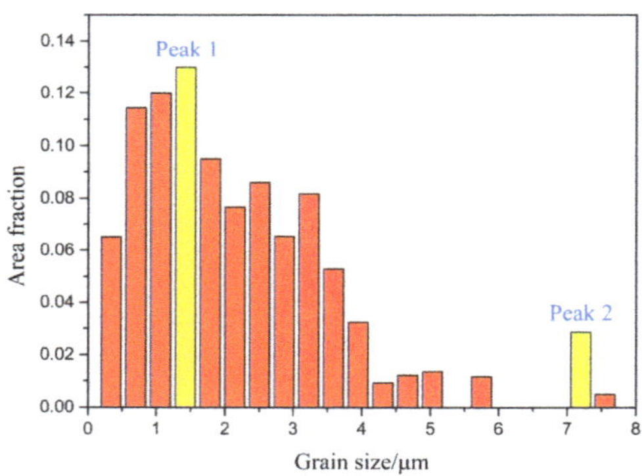

Figure 7. Grain size distribution of the fusion zone in the heat-treated joint (P = 2500 W, t = 5 ms).

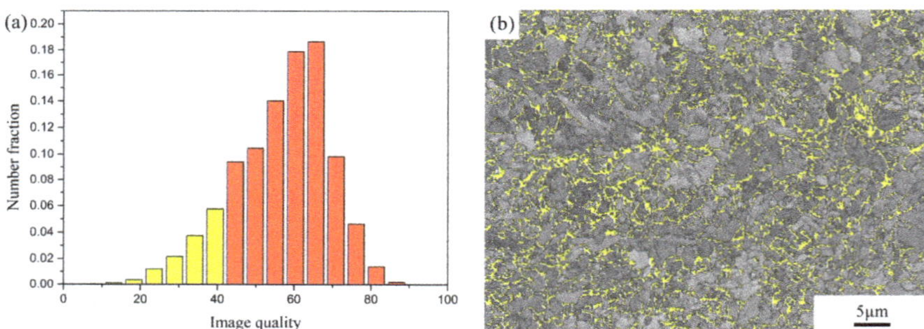

Figure 8. Fusion zone in the heat-treated joint (P = 2500 W, t = 5 ms): (**a**) Image quality in percentage showing highlighted low image quality and (**b**) Image quality map showing highlighted phases corresponding to martensite.

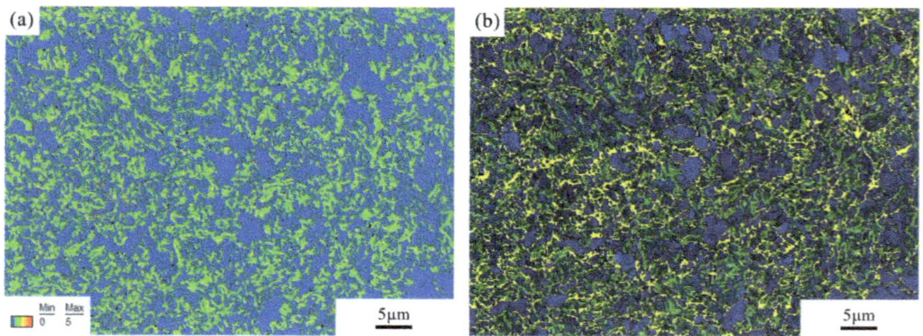

Figure 9. (**a**) Kernel average misorientation (KAM) map of the fusion zone in the heat-treated joint (P = 2500 W, t = 5 ms) and (**b**) Kernel average misorientation (KAM) map combined with image quality (IQ) map shown in Figure 8b.

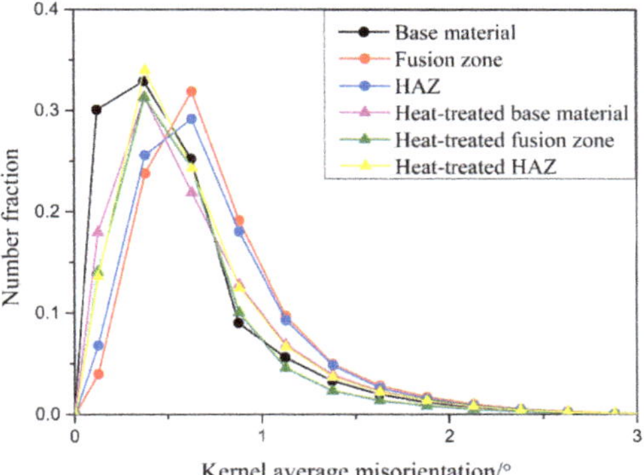

Figure 10. Distribution of kernel average misorientation (KAM) of different regions in the as-joined and heat-treated joints.

4. Discussion

With controlled welding conditions, including an optimised laser power and a short pulse duration, a fully penetrated and not overheated spot weld was obtained. The welding defects observed in the joint are mainly material loss and microvoids. The material loss is generally caused by metal evaporation and spattering. This could be improved by tuning the focal position and the divergence angle [24,25]. With a reduced power density around the keyhole aperture, the speed of the melt flow as well as the instability of the keyhole will be decreased; therefore, the evaporation and spattering phenomenon will be limited. As for the porosity defects, it has been found that porosity can be effectively inhibited at the optimum frequency and duty cycle of the employed laser pulse [26]. In addition, the keyhole stability and material evaporation also have a large effect on the generation of porosity [27]. The use of nitrogen instead of inert shielding gas was found to be effective to suppress the formation of porosity defects, due to the occurrence of the on-and-off cycle of nitrogen plasma prior to the initiation of keyhole instability [27]. All of these could be used to reduce the welding defects and thus, improve the mechanical performance of the weld in future works.

The melt pool lifetime during fusion welding is crucial for retaining the microstructure and mechanical properties of ODS steel joints. Upon welding, the oxide particles will quickly float to the top of the melt pool and, consequently, form a depleted area, leading to significantly reduced mechanical properties. With traditional "continuous" welding, the heat will accumulate in the weld. Even though techniques such as laser beam welding have a very high cooling rate, the melt pool lifetime is still too long for ODS steels. The temperature of the fusion zone usually stays above the melting point for a relatively long period of time. It is therefore difficult, if not impossible, to obtain an undisturbed microstructure, i.e., a microstructure with fine grains and dispersed nano-oxide particles. For instance, Lindau et al. [9] studied the joining of ODS Eurofer via electron beam welding, which also has a characteristic high power density. The results showed that the nano-dispersoids in the fusion zone unsurprisingly agglomerated to larger particles, causing a weak weld seam. Conversely, with the distributed pulsed spot welding procedure proposed in this study, the melt pool lifetime of each spot is reduced to the order of milliseconds, which is favourable for the retention of the microstructure and mechanical properties of the joint. It is known that the strengthening mechanisms of ODS steels are mainly based on grain boundary strengthening and dispersion strengthening [28], as neither the grain size nor the nanoparticle distribution is changed significantly after joining, and the strength of the joint will not be unduly reduced.

5. Conclusions

ODS Eurofer was welded successfully using a pulsed laser beam welding technique with a distributed pulse pattern. The welding parameters were optimised based on their effect on the microstructure. The microstructure of the joint in the as-joined and heat-treated conditions was investigated in detail. The main conclusions are as follows.

1. A full penetration is necessary to obtain an open keyhole condition without trapped gas. With a laser power of 2500 W, a pulse duration of at least 3 ms is needed for a 1 mm thick sample. A longer duration may lead to excess material loss, which can be compensated by the addition of filler materials.
2. In the as-joined condition, the fusion zone consists of elongated grains, while the HAZ has refined grains compared to the base material. $M_{23}C_6$ carbides are observed in the microstructure and found to be preferentially located at the grain boundaries. The nanoprecipitates are retained in the fusion zone, although they are not homogenously distributed in the steel matrix. Characterisations using small angle neutron scattering or small angle X-ray scattering could be carried out in the future to better understand the effect of laser beam welding on nanocluster behaviour.
3. EBSD results reveal that the joined material has a bimodal and dual phase microstructure. Martensite grains show a lower image quality and a higher misorientation compared to ferrite grains. The microstructure is generally recovered and homogenous after the heat treatment, which is beneficial for the mechanical properties. Our study shows that pulsed laser beam welding is an effective method for the joining of ODS Eurofer steel.

Author Contributions: Conceptualization, I.R.; methodology, I.R.; software, J.F.; validation, J.F.; formal analysis, J.F.; investigation, J.F.; writing—original draft preparation, J.F.; writing—review and editing, I.R. and M.H.; visualization, J.F. and M.H.; supervision, M.H.; project administration, M.H.; funding acquisition, M.H. All authors have read and agreed to the published version of the manuscript.

Funding: This research was carried out under project number T16010f in the framework of the Partnership Program of the Materials innovation institute M2i (www.m2i.nl, accessed on 28 April 2021) and the Netherlands Organisation for Scientific Research (www.nwo.nl, accessed on 28 April 2021).

Data Availability Statement: The raw/processed data required to reproduce these findings cannot be shared at this time as the data also form part of an ongoing study.

Conflicts of Interest: The authors declare no conflict of interest.

References

1. Zinkle, S.; Boutard, J.; Hoelzer, D.; Kimura, A.; Lindau, R.; Odette, G.; Rieth, M.; Tan, L.; Tanigawa, H. Development of next generation tempered and ODS reduced activation ferritic/martensitic steels for fusion energy applications. *Nucl. Fusion* **2017**, *57*, 092005. [CrossRef]
2. Dash, M.K.; Mythili, R.; Ravi, R.; Sakthivel, T.; Dasgupta, A.; Saroja, S.; Bakshi, S.R. Microstructure and mechanical properties of oxide dispersion strengthened 18Cr-ferritic steel consolidated by spark plasma sintering. *Mater. Sci. Eng. A* **2018**, *736*, 137–147. [CrossRef]
3. Baker, B.; Brewer, L. Joining of oxide dispersion strengthened steels for advanced reactors. *JOM* **2014**, *66*, 2442–2457. [CrossRef]
4. Fu, J.; Brouwer, J.; Richardson, I.; Hermans, M. Joining of oxide dispersion strengthened Eurofer steel via spark plasma sintering. *Mater. Lett.* **2019**, *256*, 126670. [CrossRef]
5. Noh, S.; Kasada, R.; Kimura, A. Solid-state diffusion bonding of high-Cr ODS ferritic steel. *Acta Mater.* **2011**, *59*, 3196–3204. [CrossRef]
6. Chen, C.-L.; Tatlock, G.; Jones, A. Microstructural evolution in friction stir welding of nanostructured ODS alloys. *J. Alloys Compd.* **2010**, *504*, S460–S466. [CrossRef]
7. Zhao, K.; Liu, Y.; Huang, L.; Liu, B.; He, Y. Diffusion bonding of Ti-45Al-7Nb-0.3 W alloy by spark plasma sintering. *J. Mater. Process. Technol.* **2016**, *230*, 272–279. [CrossRef]
8. Chen, Z.; Cui, S. On the forming mechanism of banded structures in aluminium alloy friction stir welds. *Scripta Mater.* **2008**, *58*, 417–420. [CrossRef]

9. Lindau, R.; Klimenkov, M.; Jäntsch, U.; Möslang, A.; Commin, L. Mechanical and microstructural characterization of electron beam welded reduced activation oxide dispersion strengthened–Eurofer steel. *J. Nucl. Mater.* **2011**, *416*, 22–29. [CrossRef]
10. Kelly, T. Pulsed YAG laser welding of ODS alloys. In *AIP Conference Proceedings*; American Institute of Physics: College Park, MD, USA, 1979; pp. 215–220.
11. Molian, P.; Yang, Y.; Patnaik, P. Laser welding of oxide dispersion-strengthened alloy MA754. *J. Mater. Sci.* **1992**, *27*, 2687–2694. [CrossRef]
12. Lemmen, H.; Sudmeijer, K.; Richardson, I.; Van der Zwaag, S. Laser beam welding of an Oxide Dispersion Strengthened super alloy. *J. Mater. Sci.* **2007**, *42*, 5286–5295. [CrossRef]
13. Liang, S.; Lei, Y.; Zhu, Q. The filler powders laser welding of ODS ferritic steels. *J. Nucl. Mater.* **2015**, *456*, 206–210. [CrossRef]
14. Fu, J.; Brouwer, J.; Richardson, I.; Hermans, M. Effect of mechanical alloying and spark plasma sintering on the microstructure and mechanical properties of ODS Eurofer. *Mater. Des.* **2019**, *177*, 107849. [CrossRef]
15. De, A.; Walsh, C.; Maiti, S.; Bhadeshia, H. Prediction of cooling rate and microstructure in laser spot welds. *Sci. Technol. Weld. Join.* **2003**, *8*, 391–399. [CrossRef]
16. Yamamoto, M.; Ukai, S.; Hayashi, S.; Kaito, T.; Ohtsuka, S. Formation of residual ferrite in 9Cr-ODS ferritic steels. *Mater. Sci. Eng. A* **2010**, *527*, 4418–4423. [CrossRef]
17. Zilnyk, K.D.; Sandim, H.R.Z.; Bolmaro, R.E.; Lindau, R.; Möslang, A.; Kostka, A.; Raabe, D. Long-term microstructural stability of oxide-dispersion strengthened Eurofer steel annealed at 800 °C. *J. Nucl. Mater.* **2014**, *448*, 33–42. [CrossRef]
18. Sallez, N.; Donnadieu, P.; Courtois-Manara, E.; Chassaing, D.; Kübel, C.; Delabrouille, F.; Blat-Yrieix, M.; de Carlan, Y.; Bréchet, Y. On ball-milled ODS ferritic steel recrystallization: From as-milled powder particles to consolidated state. *J. Mater. Sci.* **2015**, *50*, 2202–2217. [CrossRef]
19. Sallez, N.; Boulnat, X.; Borbély, A.; Béchade, J.L.; Fabrègue, D.; Perez, M.; de Carlan, Y.; Hennet, L.; Mocuta, C.; Thiaudière, D.; et al. In situ characterization of microstructural instabilities: Recovery, recrystallization and abnormal growth in nanoreinforced steel powder. *Acta Mater.* **2015**, *87*, 377–389. [CrossRef]
20. Hilger, I.; Bergner, F.; Weißgärber, T. Bimodal Grain Size Distribution of Nanostructured Ferritic ODS Fe-Cr Alloys. *J. Am. Ceram. Soc.* **2015**, *98*, 3576–3581. [CrossRef]
21. Odette, G.; Alinger, M.; Wirth, B. Recent developments in irradiation-resistant steels. *Annu. Rev. Mater. Res.* **2008**, *38*, 471–503. [CrossRef]
22. Tarasiuk, J.; Gerber, P.; Bacroix, B. Estimation of recrystallized volume fraction from EBSD data. *Acta Mater.* **2002**, *50*, 1467–1477. [CrossRef]
23. Laureys, A.; Pinson, M.; Depover, T.; Petrov, R.; Verbeken, K. EBSD characterization of hydrogen induced blisters and internal cracks in TRIP-assisted steel. *Mater. Charact.* **2020**, *159*, 110029. [CrossRef]
24. Kawahito, Y.; Nakada, K.; Uemura, Y.; Mizutani, M.; Nishimoto, K.; Kawakami, H.; Katayama, S. Relationship between melt flows based on three-dimensional X-ray transmission in situ observation and spatter reduction by angle of incidence and defocussing distance in high-power laser welding of stainless steel. *Weld. Int.* **2018**, *32*, 485–496. [CrossRef]
25. Weberpals, J.-P.; Krueger, P.; Berger, P.; Graf, T. Understanding the influence of the focal position in laser welding on spatter reduction. In Proceedings of the International Congress on Applications of Lasers & Electro-Optics, Orlando, FL, USA, 23–27 October 2011; pp. 159–168.
26. Dowden, J.; Kapadia, P. A mathematical investigation of the penetration depth in keyhole welding with continuous CO2 lasers. *J. Phys. D Appl. Phys.* **1995**, *28*, 2252. [CrossRef]
27. Matsunawa, A.; Mizutani, M.; Katayama, S.; Seto, N. Porosity formation mechanism and its prevention in laser welding. *Weld. Int.* **2003**, *17*, 431–437. [CrossRef]
28. Zhou, X.; Liu, Y.; Yu, L.; Ma, Z.; Guo, Q.; Huang, Y.; Li, H. Microstructure characteristic and mechanical property of transformable 9Cr-ODS steel fabricated by spark plasma sintering. *Mater. Des.* **2017**, *132*, 158–169. [CrossRef]

 micromachines

MDPI

Article

A Comparative Study on the Wettability of Unstructured and Structured $LiFePO_4$ with Nanosecond Pulsed Fiber Laser

Mulugeta Gebrekiros Berhe [1] and Dongkyoung Lee [1,2,*]

[1] Department of Future Convergence Engineering, Kongju National University, Cheonan 31080, Korea; mule1041@gmail.com

[2] Department of Mechanical and Automotive Engineering, Kongju National University, Cheonan 31080, Korea

* Correspondence: ldkkinka@kongju.ac.kr

Abstract: The wettability of electrodes increases the power and energy densities of the cells of lithium-ion batteries, which is vital to improving their electrochemical performance. Numerous studies in the past have attempted to explain the effect of electrolyte and calendering on wettability. In this work, the wettability behavior of structured and unstructured $LiFePO_4$ electrodes was studied. Firstly, the wettability morphology of the structured electrode was analyzed, and the electrode geometry was quantified in terms of ablation top and bottom width, ablation depth, and aspect ratio. From the result of the geometry analysis, the minimum measured values of aspect ratio and ablation depth were used as structured electrodes. Laser structuring with pitch distances of 112 μm, 224 μm, and 448 μm was applied. Secondly, the wettability of the electrodes was measured mainly by total wetting time and electrolyte spreading area. This study demonstrates that the laser-based structuring of the electrode increases the electrochemically active surface area of the electrode. The electrode structured with 112 μm pitch distance exhibited the fastest wetting at a time of 13.5 s. However, the unstructured electrode exhibited full wetting at a time of 84 s.

Keywords: wettability; electrodes; laser structuring; spread area; electrolyte; wetting time

Citation: Berhe, M.G.; Lee, D. A Comparative Study on the Wettability of Unstructured and Structured $LiFePO_4$ with Nanosecond Pulsed Fiber Laser. *Micromachines* **2021**, *12*, 582. https://doi.org/10.3390/mi12050582

Academic Editors: Yu Huang and Congyi Wu

Received: 8 May 2021
Accepted: 19 May 2021
Published: 20 May 2021

1. Introduction

The energy demand of daily human activities is growing fast, causing climate change as a result of greenhouse gas emissions. The automotive industry is labeled the primary contributor to this climate change [1]. Automotive battery technology is essential for the production of efficient, sustainable, safe, stable, and low emission hybrid and electric vehicles to overcome the above-mentioned challenges [2]. In contrast to lead-acid batteries and nickel-cadmium batteries, lithium-ion batteries have high energy and power density, a lighter weight, are nontoxic, and have a long cycle life [2–6].

The main components of a lithium-ion battery are the anode, cathode, separator, and electrolyte. The anode is the negative electrode, which usually uses graphite as an active electrode material. The active electrode material is coated on a copper current collector foil. The cathode, meanwhile, is made of lithium metal oxide that coats the aluminum current collector. During the battery charging process, lithium ions and electrons flow from the cathode to the anode through an external circuit and separator, and vice-versa during discharging [7–9]. As mentioned above, the active material of the battery are the main elements that determine the performance of the battery. $LiFePO_4$ (lithium iron phosphate) is one of the most widely used cathode materials in energy storage. It offers excellent electrochemical performance because of its low cost and thermal stability, as well as the fact that it is environmentally safe compared to other cathode materials [10].

During the past few years, technologies that have been used for cutting, welding, annealing, structuring, and surface processing of lithium-ion batteries have been replaced with a new and simple method using a laser, as the laser is a non-contact process and can perform precision processes. 2D electrode designs exhibit long ion transport distances, power losses, and interelectrode ohmic resistance, which results in the restricted performance of the battery by minimizing energy and power densities. To overcome these disadvantages, researchers have come up with the idea of the 3D electrode. 3D electrode architectures are obtained by removing the active material partially and increasing the areal energy capacity to overcome the above-mentioned limitations and mechanical degradation during battery operation [9,11,12]. A battery's performance is mainly determined by its electrode performance, and laser structuring of an electrode is mainly aimed at improving this.

Wettability is one of the main properties that greatly affects the electrode's performance, as insufficient and non-uniform wettability of an electrode results in poor capacity, production failure, reduced lifetime, and degradation [9,13–15]. In the past, numerous studies have attempted to investigate electrode surface characteristics and their effect on wettability. The first study analyzed the effect of electrode calendering and found that calendering of the electrode changes the pore structure, thereby improving the wetting behavior [11,16]. The wetting behavior of laser-induced electrode surface structures such as holes, cone, grid, and line structures were studied, and results indicated that surface modification has a huge impact on improving wettability [12]. Other studies looked at the effect of electrolyte solution on the wettability of electrodes, and found that the wetting rate mainly depends on the viscosity and surface tension of the electrolyte solution [17]. However, most studies have focused on the effect of electrolytes with different compositions on wettability, and there is a lack of study on the effect of laser processed electrode surface on wettability. Hence, this study aims to investigate the effect of the electrode surface microstructure. In addition, a wettability test is conducted to compare the wetting behavior of micro-structured electrodes.

In the present experiment, the effect of laser power on the structured electrode was observed in terms of ablation width, ablation depth, and aspect ratio. The structured electrode microstructure was analyzed using SEM. In addition, the wettability of unstructured and structured electrodes was analyzed in terms of wetting time and spread area. This paper is organized as follows. First, an experimental setup, material information, laser source, and laser parameters are described. Second, the surface morphology, ablation width, ablation depth, and aspect ratio are discussed. Third, the wettability of the electrodes is explained and discussed. Finally, the concluding remarks of this study are summarized.

2. Experimental Setup

2.1. Material

For this study, a one-side coated $LiFePO_4$ electrode with aluminum as a current collector was used. The slurry was prepared by mixing active material $LiFePO_4$, super P as a conducting agent, and polyvinylidene fluoride as a binder with a mass fraction of 8:1:1, respectively. Active material with a thickness of 70 μm was coated on 20 μm of aluminum foil. After calendering, the electrode total thickness was reduced to 76 μm. At the same time, the active electrode material thickness was also reduced to 56 μm. A $LiPF_6$ liquid electrolyte composed of a solvent mixture of ethylene carbonate (EC) and ethyl methyl carbonate (EMC) with a volume ratio of 3:7 was used for the wettability test. The measurements were carried out as static experiments under ambient air using a drop volume of 0.01 mL [18].

2.2. Laser Source

A Ytterbium nanosecond pulsed fiber laser (YLPM-1-4 × 200-20-20, IPG, NY, New York, USA), shown in Figure 1, was used as a laser source for this experiment. It features a maximum average laser power of 20 W, an emission wavelength of 1064 nm, and a maximum pulse duration of 200 ns. The laser source also had a beam quality factor (M^2) of 1.5, a collimated beam diameter of 7.2 mm, and a spot diameter of 30 μm.

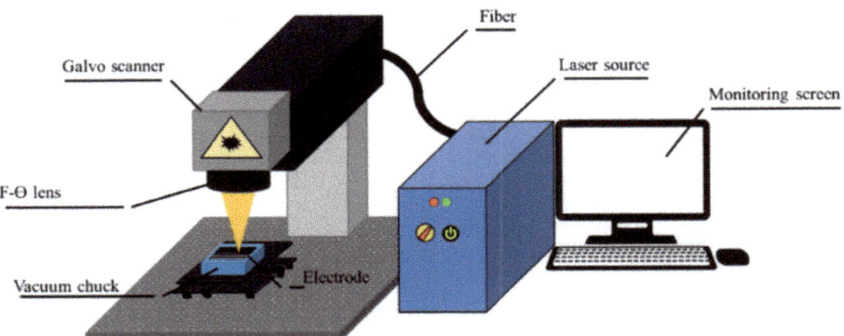

Figure 1. Set up of Ytterbium nanosecond Pulsed Fiber Laser used for laser structuring.

2.3. Experiment Procedure

The $LiFePO_4$ electrode was structured with a two-pass method using a laser power of 1–4.6 W (at 0.4 W intervals). The electrode structure with the minimum groove size was chosen for further laser structuring with three different pitch distances of 112 μm, 224 μm, and 448 μm. To evaluate the wettability of structured and unstructured electrodes, electrolyte spread area and total wetting time were analyzed. The effect of surface microstructure on the wetting behavior of both a structured and an unstructured electrode was investigated. The experiment parameters employed for this study are indicated in Table 1.

Table 1. Experiment parameters.

Laser Parameters	
Laser power(W)	1~4.6 W
Wavelength	1064 nm
Pulse duration	4 ns
Pulse repetition rate	500 kHz
Scanning speed	500 mm/s
Number of passes	2 passes
Pitch distances	112 μm, 224 μm and 448 μm
Working distance	189 mm

3. Result and Discussion

3.1. Morphology

A scanning Electron Microscope (SEM) was employed to observe the top and cross-section macrostructure of the electrode structures as shown in Figure 2. From both top and cross-section views, the formation of a clear groove was observed from the minimum laser power of 1 W. The spherical solidified molten active electrode material of approximately 5–8 μm in size was observed from the laser power of 1 W. The size of these melt formations increases as the laser power increases. Under higher laser powers, observation reveals an increase in molten active electrode materials extant on the top surface.

3.2. Ablation Width, Ablation Depth, and Aspect Ratio

The ablation top and bottom widths after being subjected to all laser powers in are shown in Figure 3a. When the laser power increases, both the ablation top and bottom width increase. The measured ablation top width ranges from 35 μm to 60 μm when using laser powers from 1 W to 4.6 W. Meanwhile, the ablation bottom width varies from 8 μm to 35 μm over the same power range. From the laser power of 1 W to 3.8 W, the ablation top width increases rapidly. On the other hand, the bottom width increases significantly at the laser power ranging from 3 W to 4.6 W.

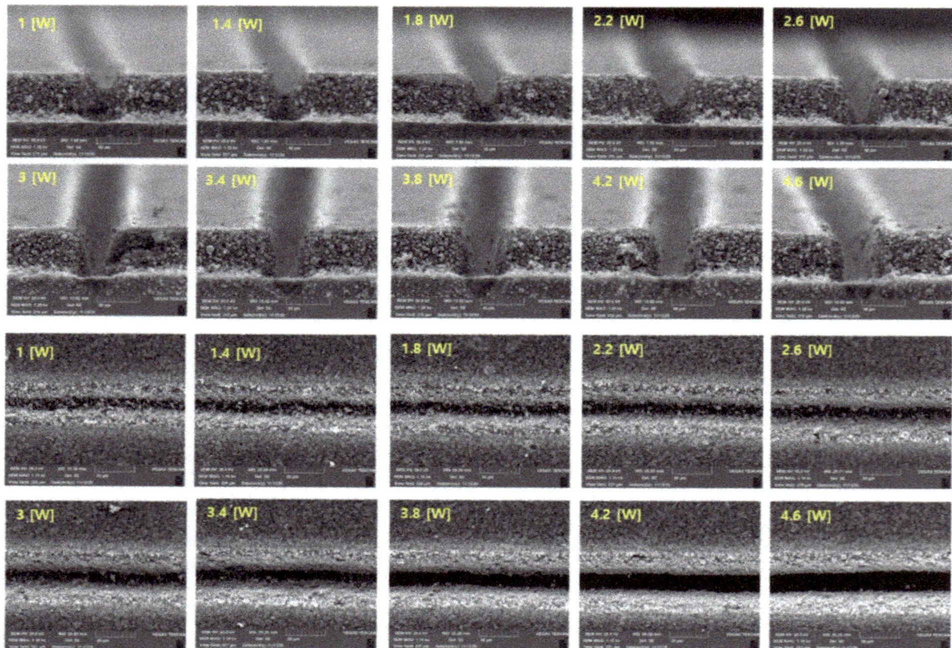

Figure 2. SEM image of the electrode after the laser structuring with two-pass.

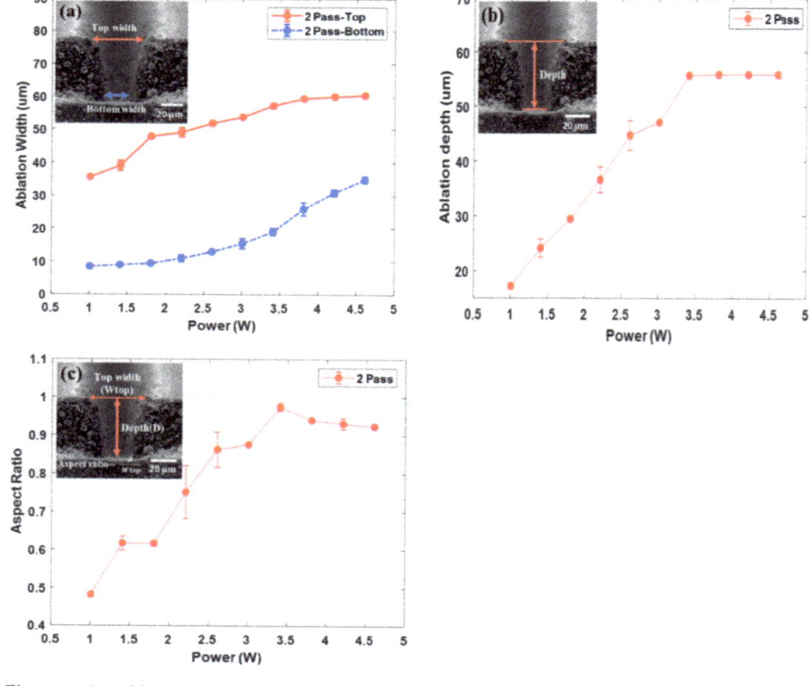

Figure 3. (**a**) Ablation width, (**b**) ablation depth, and (**c**) aspect ratio.

Figure 3b,c show the ablation depth and aspect ratio of the laser structured electrode. An ablation depth of 17 µm is formed at the laser power of 1 W. The ablation depth increases as the laser power increases. The maximum ablation depth is obtained at the laser power of 3.4 W, at which point the aluminum foil gets exposed. The measured maximum ablation depth is 56 µm. The ablation depth remains stable for the laser powers from 3.4 W to 4.6 W because the active electrode material is completely removed.

The reason ablation depth does not change after the active electrode material is fully ablated indicates that the fluence of the laser beam cannot exceed the ablation threshold of the aluminum foil [17]. The laser powers from 3.4 W to 4.6 W contribute to the increase of the ablation top and bottom width, but do not affect the ablation depth. The ablation depth has reached its maximum possible value at the laser power of 3.4 W. The aspect ratio calculated using the proportion of ablation depth to ablation top width is shown in Figure 3c. The aspect ratio is a very important parameter. The maximum aspect ratio results in creating an increased active electrode material surface area for better electrochemical reaction and enhanced energy density of the battery [18]. The maximum average aspect ratio was found to be 0.974, and was observed when the laser power of 3.4 W was used. The maximum aspect ratio is obtained when the active electrode material is completely removed. Moreover, the aspect ratio decreases as the laser power increase from 3.4 W. The reason is that the ablation width keeps increasing while the ablation depth remains unchanged. As the result, the aspect ratio declines. To clearly understand the wettability phenomenon between structured and unstructured from the start, the groove with the smallest depth and aspect ratio measurements was chosen. The groove formed at the laser power of 1 W is structured again for further wettability tests. It was laser structured using three different pitch distances of 112 µm, 224 µm, and 448 µm.

3.3. Wettability

Wetting property is dependent on drop spreading, electrolyte imbibition, and wicking in the surface [16]. The electrolyte diffusion rate of a battery cell contributes significantly to its capacity, safety, and life cycle. Moreover, the wetted volume of drop and wetting time are the main factors of electrolyte diffusion [16]. The structures formed on the surfaces of the electrodes result in a significant increase in the electrode surface area. According to Wenzel, the wettability of the structured electrodes can be improved by modifying the electrode's surface (i.e., increasing surface roughness) [16,17,19,20]. In this study, electrodes wettability was compared using the total wetting time and spread area at specific intervals. The unstructured electrode and structured electrodes at a laser power of 1 W with different pitch distances are shown in Figure 4.

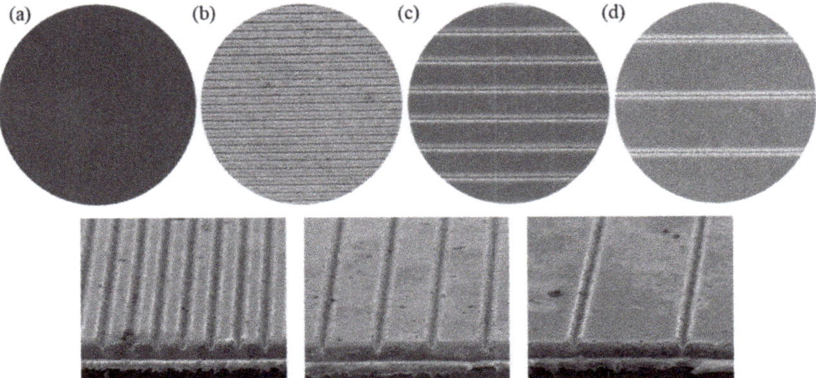

Figure 4. (**a**) Unstructured, (**b**) pitch distance of 112 µm, (**c**) pitch distance of 224 µm, and (**d**) pitch distance of 448 µm.

Wettability is evaluated in terms of the wetting time and spreading area. The total wetting time is the time required for the electrolyte to fully penetrate the electrode. The spreading area of the electrolyte is the area covered by the liquid electrolyte while diffusing over the electrode surface before fully penetrating the electrode. Additionally, the spread area measurement starts from the initial drop to the spread area where a faintly visible contact angle between the electrode surface and electrolyte drop can be obtained at different times, t.

Figure 5 shows the electrolyte spread area at different wetting times and expresses the quantitative electrolyte covered area. Additional pictures for the unstructured electrode at 30, 50, 70, and 90 s were also taken. The total time required for the electrolyte to fully penetrate the unstructured electrode was 84 s. In Figure 6, the total wetting time for the electrode structured with 112 μm pitch was 13.5 s, the electrode structured with 224 μm was 23.5 s, and the electrode structured with 448 μm was 33 s. The wetting time difference between the structured electrodes is approximately 10 s. However, the unstructured electrode took a long time compared to the structured electrodes.

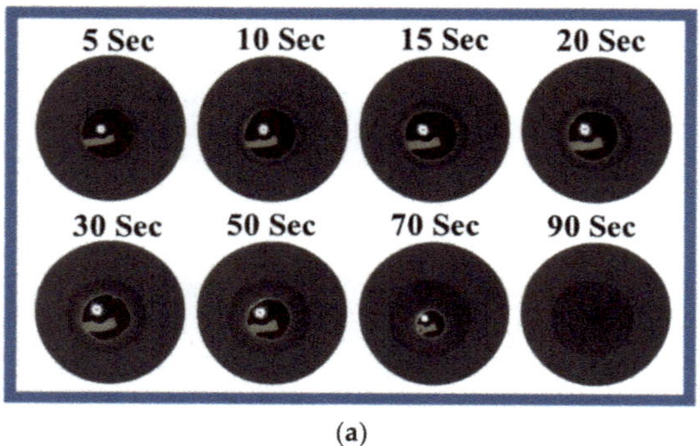

(a)

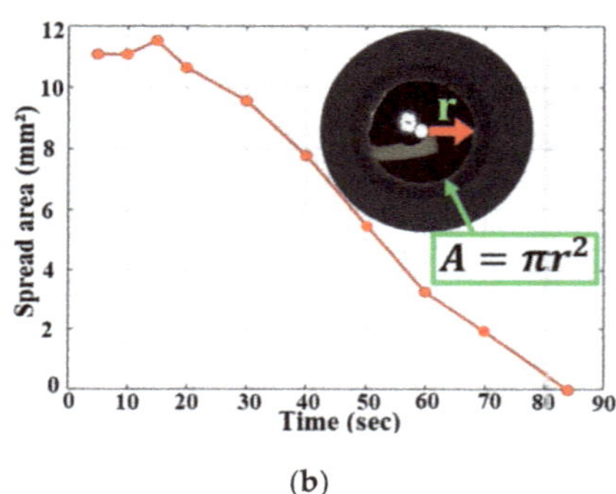

(b)

Figure 5. (**a**) Images of an unstructured electrode at different points in time. (**b**) Electrolyte spread over the unstructured electrode with time.

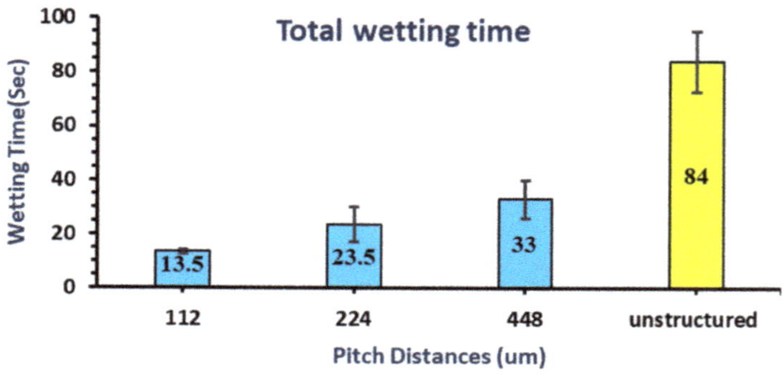

Figure 6. Total wetting time of all electrodes.

Taken together, these results indicate that the wetting time of structured electrodes increases when the pitch distance between grooves increases. Electrodes with small pitch distances have less wetting time than those with bigger pitch distances. This is due to an increased active surface area caused by smaller pitch distances.

Moreover, the wettability is also evaluated by analyzing the wetted surface area after 5, 10, 15, and 20 s, respectively. The unstructured electrode showed different wettability phenomena, unlike the structured electrode. The electrolyte droplet is stable on the surface and does not spread. Rather, it wets uniformly in all directions in a circular shape. In the unstructured electrode, the drop is radially symmetrical around the center which is due to the uniform pore distribution of the electrode. In addition, the spread area starts falling with time without showing much difference from the initial drop. The drop spreads over a maximum area of 15 mm^2 out of the total area of 94 mm^2. The covered area while spreading can be seen in Figure 7b with time until it becomes fully wet. The images of the structured electrode recorded at 5 s indicate that the spread area of the liquid on the structured electrode with 112 μm pitch distance is larger than the two other structured electrodes. It is also seen that the electrolyte spread area decreases while increasing the pitch distance between grooves.

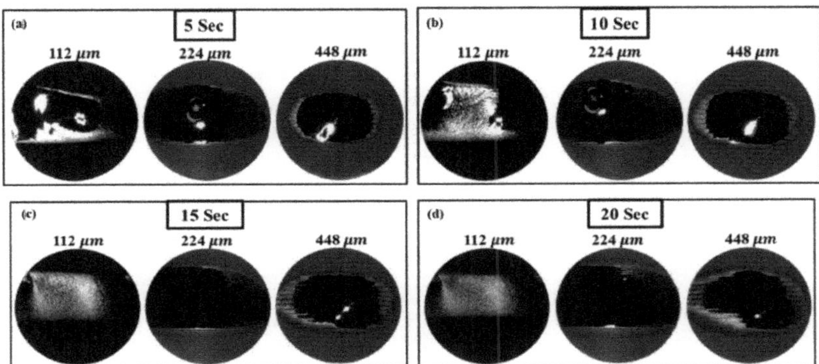

Figure 7. Images of the laser structured electrode with pitch distances of 112 μm, 224 μm, and 448 μm, respectively at the time of (**a**) 5 s, (**b**) 10 s, (**c**) 15 s, and (**d**) 20 s.

In comparison with the electrodes with 112 μm and 224 μm pitch distances, the electrode with a 448 μm pitch distance seems to have an elliptical shape spread area and is stable in a fixed zone. After 10 s, more than half of the electrolyte drop has wetted in the electrode with a 112 μm pitch distance structure while the other two are still spreading. After 15 and 20 s, the entire drop placed on the electrode structure with a 112 μm pitch distance is fully spread, and has even started drying while the others are in the process of wetting. The electrolyte spreads along lines parallel to the direction of the line structures. Unlike unstructured electrodes, the spread area of the structured electrodes, in general, is not radially symmetrical because the groove is formed, and the pore distribution is affected by laser structuring. It is seen that the electrolyte spreads and wets quickly on the structured electrode with the smallest pitch distance compared to bigger pitch distances. This implies that the active electrode material surface area increases as the pitch distance gets smaller.

4. Conclusions

In this study, $LiFePO_4$ electrodes were structured using a nanosecond pulsed fiber laser via two different number passes and three different pitch distances. The wettability of the unstructured electrode and structured electrodes was discussed comparatively. The wettability of the electrodes was mainly measured and analyzed by the electrolyte's ability to penetrate quickly and its spread area. The key observations of this study can be summarized as follows:

1. The ablation top and bottom widths increase as the laser power increases. Similarly, ablation depth also increases as the laser power increases. The aspect ratio increases as the laser power increases until it starts falling at the laser power of 3.4 W, where the depth reaches its maximum limit while width keeps increasing.

2. The maximum ablation top width was measured to be 60 μm at 4.6 W, and the maximum ablation depth was 56 μm starting from a laser power of 3.4. The minimum measurements were found when using a laser power of 1 W. 35 μm and 17 μm were the minimum measured values of ablation width and depth, respectively. The maximum aspect ratio of 0.974 was obtained using a laser power of 3.4 W.
3. The wettability test results obtained by comparing the unstructured and structured electrodes showed that the surface microstructure influences the wetting performance of the electrode. It was observed that the electrode structured with a small pitch distance exhibits fast wetting and spreading behavior. It has also been noticed that the groove structures play an important role in guiding the electrolyte flow direction. The unstructured electrode shows very slow wetting and a small spread area compared to the structured electrodes.
4. This study demonstrated that the electrolyte drop chooses a differently shaped spread pattern to minimize the energy required for spreading depending on the pattern and geometry of the structured electrodes [18].

Further study may focus on investigating the wettability of laser-induced electrodes and their effect on battery performance using better techniques.

Author Contributions: D.L. and M.G.B. conceived and designed the experiments; D.L. and M.G.B. performed the experiments; D.L. and M.G.B. analyzed the data; D.L. and M.G.B. wrote the paper. All authors have read and agreed to the published version of the manuscript.

Funding: The research described herein was sponsored by the National Research Foundation of Korea (NRF) grant funded by the Korean government (MSIP; Ministry of Science, ICT & Future Planning) (No. 2019R1A2C1089644) and by the KOREA INNOVATION FOUNDATION grant funded by the Korean government (MSIP) (No. 2020-DD-SB-0159). The opinions expressed in this paper are those of the authors and do not necessarily reflect the views of the sponsors.

Conflicts of Interest: The authors declare no conflict of interest.

References

1. Lutey, A.H.; Fortunato, A.; Ascari, A.; Carmignato, S.; Leone, C. Laser cutting of lithium iron phosphate battery electrodes: Characterization of process efficiency and quality. *Opt. Laser Technol.* **2015**, *65*, 164–174. [CrossRef]
2. Lee, D.; Patwa, R.; Herfurth, H.; Mazumder, J. Parameter optimization for high speed remote laser cutting of electrodes for lithium-ion batteries. *J. Laser Appl.* **2016**, *28*, 022006. [CrossRef]
3. Liu, X.; Li, L.; Li, G. Partial surface phase transformation of Li3VO4 that enables superior rate performance and fast lithium-ion storage. *Tungsten* **2019**, *1*, 276–286. [CrossRef]
4. Chen, K.; Yin, S.; Xue, D. Active La–Nb–O compounds for fast lithium-ion energy storage. *Tungsten* **2019**, *1*, 287–296. [CrossRef]
5. Lee, D. Investigation of physical phenomena and cutting efficiency for laser cutting on anode for Li-Ion batteries. *Appl. Sci.* **2018**, *8*, 266. [CrossRef]
6. Zubi, G.; Dufo-López, R.; Carvalho, M.; Pasaoglu, G. The lithium-ion battery: State of the art and future perspectives. *Renew. Sustain. Energy Rev.* **2018**, *89*, 292–308. [CrossRef]
7. Herfurth, H.J.; Patwa, R.; Pantsar, H. Laser cutting of electrodes for advanced batteries. In Proceedings of the LPM2010—The 11th International Symposium on Laser Precision Microfabrication, Stuttgart, Germany, 8–10 June 2010; pp. 1–5.
8. Demir, A.G.; Previtali, B. Remote cutting of Li-ion battery electrodes with infrared and green ns-pulsed fibre lasers. *Int. J. Adv. Manuf. Technol.* **2014**, *75*, 1557–1568. [CrossRef]
9. Pfleging, W. A review of laser electrode processing for development and manufacturing of lithium-ion batteries. *Nanophotonics* **2018**, *7*, 549–573. [CrossRef]
10. Zhang, W.-J. Structure and performance of $LiFePO_4$ cathode materials: A review. *J. Power Sources* **2011**, *196*, 2962–2970. [CrossRef]
11. Davoodabadi, A.; Li, J.; Liang, Y.; Wood, D.L.; Singler, T.J.; Jin, C. Analysis of electrolyte imbibition through lithium-ion battery electrodes. *J. Power Sour.* **2019**, *424*, 193–203. [CrossRef]
12. Wu, M.-S.; Liao, T.-L.; Wang, Y.-Y.; Wan, C.-C. Assessment of the Wettability of Porous Electrodes for Lithium-Ion Batteries. *J. Appl. Electrochem.* **2004**, *34*, 797–805. [CrossRef]
13. Pfleging, W.; Mangang, M.; Zheng, Y.; Smyrek, P. Laser structuring for improved battery performance. *SPIE Newsroom* **2016**, *1*, 10–12. [CrossRef]
14. Faubert, P.; Müller, C.; Reinecke, H.; Smyrek, P.; Proell, J.; Pfleging, W. Femtosecond laser structuring of novel electrodes for 3D fuel cell design with increased reaction surface. *MRS Proc.* **2015**, *1777*, 7–13. [CrossRef]

15. Davoodabadi, A.; Li, J.; Zhou, H.; Wood, D.L.; Singler, T.J.; Jin, C. Effect of calendering and temperature on electrolyte wetting in lithium-ion battery electrodes. *J. Energy Storage* **2019**, *26*, 101034. [CrossRef]
16. Mohammadian, S.K.; Zhang, Y. Improving wettability and preventing Li-ion batteries from thermal runaway using microchannels. *Int. J. Heat Mass Transf.* **2018**, *118*, 911–918. [CrossRef]
17. Liu, T.; Wang, K.; Chen, Y.; Zhao, S.; Han, Y. Dominant role of wettability in improving the specific capacitance. *Green Energy Environ.* **2019**, *4*, 171–179. [CrossRef]
18. Proll, J.; Köhler, R.; Adelhelm, C.; Bruns, M.; Torge, M.; Heisler, S.; Przybylski, M.; Ziebert, C.; Pfleging, W. Laser modification and characterization of Li-Mn-O thin film cathodes for lithium-ion batteries. *Int. Soc. Opt. Photonics* **2011**, *7921*, 79210Q. [CrossRef]
19. Davoodabadi, A.; Li, J.; Liang, Y.; Wang, R.; Zhou, H.; Wood, D.L.; Singler, T.J.; Jin, C. Characterization of Surface Free Energy of Composite Electrodes for Lithium-Ion Batteries. *J. Electrochem. Soc.* **2018**, *165*, A2493–A2501. [CrossRef]
20. Sheng, Y. Investigation of Electrolyte Wetting in Lithium Ion Batteries: Effects of Electrode Pore Structures and Solution. Thesis, No. December 2015. Available online: http://umich.edu/~{}racelab/static/Webpublication/2015-JPS-XuningF.pdf (accessed on 15 February 2021).

MDPI
St. Alban-Anlage 66
4052 Basel
Switzerland
Tel. +41 61 683 77 34
Fax +41 61 302 89 18
www.mdpi.com

Micromachines Editorial Office
E-mail: micromachines@mdpi.com
www.mdpi.com/journal/micromachines

www.ingramcontent.com/pod-product-compliance
Lightning Source LLC
LaVergne TN
LVHW070615170726
843515LV00004B/79

* 9 7 8 3 0 3 6 5 2 7 0 3 1 *